Color Detection

Edited by Ling-Wen Zeng and Shi-Lin Cao

Published in London, United Kingdom

IntechOpen

Supporting open minds since 2005

Color Detection
http://dx.doi.org/10.5772/intechopen.74906
Edited by Ling-Wen Zeng and Shi-Lin Cao

Contributors
Volkan Kılıç, Nesrin Horzum, Mehmet Ertugrul Solmaz, Mikhail Dolomatov, Natalia Yu. Grigoryeva, Ludmila Chistyakova, Amanda Santos, Luiz Antonio D'Avila, Gisele Borges, Kissya Da Silva, Lingwen Zeng, Shilin Cao

First published in London, United Kingdom, 2020 by IntechOpen
IntechOpen is the global imprint of INTECHOPEN LIMITED, registered in England and Wales, registration number: 11086078, 7th floor, 10 Lower Thames Street, London, EC3R 6AF, United Kingdom
Printed in Croatia

British Library Cataloguing-in-Publication Data
A catalogue record for this book is available from the British Library

Additional hard and PDF copies can be obtained from orders@intechopen.com

Color Detection
Edited by Ling-Wen Zeng and Shi-Lin Cao
p. cm.
Print ISBN 978-1-83880-599-9
Online ISBN 978-1-83880-600-2
eBook (PDF) ISBN 978-1-83880-601-9

For EU product safety concerns:
IN TECH d.o.o., Prolaz Marije Krucifikse Kozulić 3, 51000 Rijeka, Croatia, info@intechopen.com or visit our website at intechopen.com.

Meet the editors

Lingwen Zeng received his PhD degree in Genetics from McMaster University, Canada, in 1993, and his postdoctoral training from the University of Chicago, USA. He has worked in three publicly traded companies (Quest Diagnostics Inc., Scios Inc., and Genetics Computer Group Inc.) in the USA as a research scientist and a project manager. Currently, he serves as a professor at the School of Food Science and Engineering, Foshan University, China, and a principal investigator at Guangzhou Institutes of Biomedicine and Health, Chinese Academy of Sciences. His research focuses on exploring novel diagnostic technologies for human diseases, food, and environmental safety.

Shilin Cao is an associate professor at the School of Food Science and Engineering, Foshan University, China. He received his PhD in Biochemical Engineering in 2016 from the South China University of Technology. From 2015 to 2016, he worked as a visiting researcher at the Department of Applied Biology and Chemical Technology, the Hong Kong Polytechnic University in Hong Kong. His research interests include enzyme catalysis in novel solvent media, immobilization of enzymes onto novel supporting materials, and application of biobased nanomaterials.

Contents

Preface

Chromogenic assays (color detection) result in colored reaction products. By detecting the color change before and after the reaction, substance concentration can be determined by the naked eye, light microscopes, or spectrophotometers. Because of their rapid, direct, specific, convenient, and sensitive features, color detection exhibits great potential in the field of quality monitoring, chemical technology, nanophysics, and clinical medicine. The intention of this book is to provide readers with a comprehensive overview of the principles, features, and applications of color detection.

In the introduction chapter, Cao and Zeng briefly introduce the application of color detection in food analysis and biochemical detections, including protein, reducing and nonreducing sugar, pH, sulfur dioxide, and phosphorus.

Mikhail Dolomatov provides new data on the color phenomenon for complex and simple substances in recent years, and introduces the effects of the relationship of physical and chemical properties and color characteristics (in color systems RGB or XYZ) of compounds (color properties principle).

Natalia Grigoryeva and Ludmila Chistyakova present several experimental approaches to study the metabolic mechanisms in single photosynthetic cells in vivo. They are accompanied by several examples of in vivo investigations. Three main CLSM tools will be discussed in detail: spectral imaging, fluorescent microscopic spectroscopy, and FRAP.

Volkan Kılıç, Nesrin Horzum, and Mehmet Ertugrul Solmaz present a review of smartphone-based colorimetric determination of chromogenic assays that has been provided on color spaces, existing color matching and detection techniques, hardware and software designs, and performance metrics that have been developed over the past few decades.

Amanda Pereira Franco dos Santos, Kissya Kropf da Silva, Gisele Alves Borges, and Luiz Antonio d'Avila review the colorimetric techniques designed for quality monitoring and the detection of adulteration in fuels, especially simple, quick, low-cost procedures with potential to be used in the relative field.

The editors wish to express their thanks to all the authors of the chapters for their valuable contributions. Thanks also go to the IntechOpen staff members responsible for the completion of this book for free visible knowledge.

Ling-Wen Zeng and Shi-Lin Cao
Department of Food Science,
Foshan University (Northern Campus),
Foshan, China

Section 1

Introduction

Chapter 1

Introductory Chapter: Color Detection

Shi-Lin Cao, Jin-Heng Hao, Wei-Dong Ou-Yang, Zi-Shi Chen, Yong-Si Lv and Ling-Wen Zeng

1. Introduction

Chromogenic assays, also named as color detection, result in colored reaction products. By detecting the color change before and after the reaction, substance concentration could be determined by the naked eye, light microscopes, and spectrophotometers.

Nowadays, a large number of food analyses and biochemical detections could be performed by color detection and were exampled as below:

Protein is one of the main ingredients of food. Bradford method, developed by Marion M. Bradford in 1976 [1], is the most widely used colorimetric method for protein detection. The anionic dyes used in this method is Coomassie brilliant blue (such as G-250, R-150, R-250, R-350). For example, the Coomassie brilliant blue G-250 exists in three forms with different colors including cationic (red), neutral (green), and anionic (blue). After binding with protein under acidic conditions, the color of Coomassie brilliant blue G-250 dye converts from red to blue, and the protein concentration is assayed. Up to now, the relative reference of the Bradford method was cited by more than 240,000 literature [1].

The reducing sugars could be assayed by using 3,5-dinitrosalicylic acid method (known as DNS method). 3,5-Dinitrosalicylic acid, an aromatic compound, was used as the test reagent and reacted with reducing sugar to form the 3-amino-5-nitrosalicylic acid, since the 3-amino-5-nitrosalicylic acid could strongly absorb light at 540 nm. The amount of the reducing sugar could be measured by spectrophotometry [2–4].

The anthrone method is another colorimetric detection method for both reducing and nonreducing sugars assay [5]. Anthrone is a tricyclic aromatic ketone. In the acidic condition, the anthrone regent reacts with sugar, resulting in yielding a blue-green color. The absorbance of the above blue-green color solution could be measured at 620 nm.

Phenol-sulfuric acid method is also a colorimetric method for the detection of carbohydrate in food [6]. While reacting with phenol and sulfuric acid, the carbohydrate sample solution to be tested became yellow-orange in color. Since the sulfuric acid converted the nonreducing sugars to reducing sugars, the total amount of carbohydrate could be detected by this method.

The pH indicator paper could be used to measure the acidity of the solution to be tested. The main components of pH indicator paper contain methyl red [pH 4.4 (red)–6.2 (yellow)], bromocresol green [pH 3.6 (yellow)–5.4 (green)], and thyme blue [pH 6.7 (yellow)–7.5 (blue)]. The pH indicator paper that dropped different acidity solution exhibit different color, by using colorimetric card the range of the acidity could be detected naked eye.

The sulfur dioxide as well as sulfite was widely used during the food processing. Sulfur dioxide and sulfite exhibit a high effect in inhibiting nonenzymatic browning in the process of food processing and could be used as preservative to inhibit mold and bacteria. Thus, it is important for the detection of the content of sulfite. The sulfite could be reactive with o-phthaldialdehyde under alkaline environment and formed dark blue complex [9]. The sulfite content could be detected by the increase of the absorbance value at 628 nm.

Jabbari [7] developed a sensitive kinetic method for the determination of sulfite (0.05–2.5 μg/ml). There is an addition reaction between methyl green and the sulfite at pH 8 and 25°C. This results in the color fading of the methyl green. Thus, by detecting the decrease in the absorbance of the dyestuff at 625 nm by the fixed time method, the concentration of the sulfite could be spectrophotometrically monitored. The result showed that the limit of the detection and the relative standard deviation are 0.05 μg/ml and 2.88%, respectively.

The phosphorus in food could be detected by phosphor vanadium molybdenum yellow spectrophotometry [8]. Under acidic condition, the phosphorus reacts with vanadium ammonium molybdate and formed a stable yellow color vanadium-molybdenum-yellow.

Apart from the above, color detection exhibits great potential in the field of quality monitoring, chemical technology, nanophysics, and clinical medicine, because of their rapid, direct, specific, convenient, and sensitive features. The intention of this book is to provide readers with a comprehensive overview for the principles, features, and applications of color detection.

Acknowledgements

The authors gratefully acknowledge the Project of Department of Education of Guangdong Province (Young Creative Talents, Natural Science, No. 2017KQNCX217), High-Level Talent Start-Up Research Project of Foshan University (GG07016), and College Students' Innovation and Entrepreneurship Training Program (National-201911847023, Guangdong Province-S201911847091, S201911847097, Foshan University-XJ2019213) for partly funding this work.

Author details

Shi-Lin Cao*, Jin-Heng Hao, Wei-Dong Ou-Yang, Zi-Shi Chen, Yong-Si Lv and Ling-Wen Zeng*
Department of Food Science, Foshan University (Northern Campus), Foshan, China

*Address all correspondence to: shilin.cao@fosu.edu.cn; zeng6@yahoo.com

References

[1] Bradford MM. A rapid and sensitive method for the quantitation of microgram quantities of protein utilizing the principle of protein-dye binding. Analytical Biochemistry. 1976;**72**(1):248-254

[2] Sumner JB, Graham VA. Dinitrosalicylic acid: A reagent for the estimation of sugar in normal and diabetic urine. Journal of Biological Chemistry. 1921;**47**(5):5-9

[3] Lide DR. Zeitschrift für Physikalische Chemie. Boca Raton, Florida: Akademische Verlagsgesellschaft; 1998. pp. 3-318

[4] Miller GL. Use of dinitrosalicylic acid reagent for determination of reducing sugar. Analytical Chemistry. 1959;**31**(3):426-428

[5] Trevelyan WE, Forrest RS, Harrison JS. Determination of yeast carbohydrates with the anthrone reagent. Nature. 1952;**170**(期缺失): 626-627

[6] Masuko T, Minami A, Iwasaki N, Majima T, Nishimura S, Lee YC. Carbohydrate analysis by a phenol-sulfuric acid method in microplate format. Analytical Biochemistry. 2005;**339**(1):69-72

[7] Jabbari A, Shamsipur M. Kinetic spectrophotometric determination of traces of sulfite based on its additional reaction with methyl green. Microchemical Journal. 1993;**48**(3):349-355

[8] GB5009.87-2016. Food Safety National Standard-Determination of Phosphorus in Food

[9] Abdel-latif MS. New Spectrophotometric Method for Sulfite Determination. Analytical Letters. 1994;**27**(13):2601-2614

Section 2

The Recent Progress of Color Detection

Chapter 2

New Results in the Theory and Practical Application of Color

Mikhail Dolomatov

Abstract

The chapter provides new data on color phenomenon that we have discovered in recent years for complex and simple substances. For example, these are the effects of the relationship of physical and chemical properties and color characteristics (in color systems RGB or XYZ) of compounds (color properties principle). In particular, the effects of conjugation of ionization potential and electron affinity of light-absorbing molecules in the visible region were found. The theoretical explanation of these effects from the position of quantum theory is given. Using mathematical statistics, it is shown that the effects can be used for practical measurement of various properties of complex multicomponent substances. The results indicate the practical use of these effects in chemical technology, nanophysics, and clinical medicine.

Keywords: absorption spectrum, color characteristics, color systems, RGB, XYZ, physical properties, chemical properties, quantum theory, spectra of blood

1. Introduction

Even the Arab alchemists, the first chemists of the seventeenth and eighteenth centuries, talked about the relationship between color change and changes in the physical properties of various substances [1]. For example, the temperature of substances and their chemical transformations were estimated approximately by color. Here, new data on color phenomenon that we have discovered in recent years for complex and simple substances is provided. For example, these are the effects of the relationship of physical and chemical properties and color characteristics of compounds ("color properties principle") [2–10]. Color characteristics were measured by standard methods in colorimetric systems RGB or XYZ [11–13]. In particular, the effects of the relationship between the vertical ionization potentials and the electron affinity of light-absorbing molecules in the visible region were found [14–18]. The results indicate the practical use of these effects in chemical technology and nanophysics. We assume that the cause of these phenomena is quantum entanglement and strong correlation of electron states [19]. We established new physical effects between spectral densities (integral absorption, reflection, and transmission characteristics) with ionization potential and electron affinity [2–4, 15, 18–21]. We propose to use these effects in determining the energies of electronic states. Methods for determination of IP and EA for molecules and organic semiconductors have been developed. We propose to use these effects in determining the energies of electronic states. In addition, the color characteristics of biological fluids were investigated. In addition, we have determined the averaged color characteristics of the electromagnetic spectrum for aqueous solutions of hemolyzed blood, plasma, and serum from 100 donors and 95 patients with different diagnoses and different severities of their conditions. From the averaged absorption spectra, we calculated the color characteristics of the hemolyzed blood, plasma, and serum from the donors and patients by the standard CIE procedure. The blood is considered as a single, indivisible light-absorbing system in studying complex biological specimens. We studied the relationship between the color characteristics of human blood in normal and pathological conditions [22–26]. Let us consider these aspects in more detail.

2. Estimation of physicochemical properties of complex systems according to color characteristics

We have discovered new optical effects of the relationship between the physicochemical properties and color characteristics for very complex chemical systems [2–4]. In particular, the dependencies between the properties and color characteristics of multicomponent hydrocarbon systems are investigated. Dependencies between color coordinates (luminosity) and various physical and chemical characteristics of these substances are established. All results are confirmed by statistical data processing. The dependence of the properties on the CCs is linear (the law "color-properties"):

$$Z = B_0 + B_1 \cdot q \tag{1}$$

where Z is one of the physical or chemical properties, q is the one of the color characteristics of the substance (e.g., color coordinates X_j, Y_j, Z_j in the XYZ system or R_j, G_j, B_j in the RGB system; or chromaticity coordinates x_j, y_j, z_j in the XYZ system or trichromatic coordinates $\mathrm{r_j}$, $\mathrm{g_j}$, $\mathrm{b_j}$ in the RGB system; j, standard light source A, B, C, or D), and B_0, B_1 are the empirical constants dependent on the type of the source and the class of researched substances and dimensional properties.

Color coordinates of (X, Y, Z), coordinates of chromaticity (x, y, z), hue (λ), and luminosity (L) have been taken as color characteristics [11–13] . CCs of multicomponent hydrocarbonic systems have been determined by the technique of the International Committee on Illumination (Commission Internationale de l'Eclairage, CIE) [11] for four standard sources (illuminants) A, B, C, and D_{65}. The technique, corrected for optically transparent [13] medium, has been used. Electron absorption spectra of multicomponent hydrocarbon systems have been determined in toluene solutions in the range of 380–780 nm with the use of automatic spectrophotometer.

The CCs were defined on the methods CIE [12, 13] in the revised version to optical transparent solutions via the transparent coefficients—$\tau(\lambda)$. The color properties were calculated from formulas (2)–(7):

$$\begin{aligned} X &= \sum_{380}^{780} E(\lambda)\tau(\lambda)\overline{x}(\lambda)\Delta\lambda \\ Y &= \sum_{380}^{780} E(\lambda)\tau(\lambda)\overline{y}(\lambda)\Delta\lambda \\ Z &= \sum_{380}^{780} E(\lambda)\tau(\lambda)\overline{z}(\lambda)\Delta\lambda \end{aligned} \tag{2}$$

where X, Y, and Z are the tristimulus values for system CIE; $E(\lambda)$ *is* the spectral power distribution for the spectrum of emission source; $\overline{x}$, $\overline{y}$, and $\overline{z}$ are three color-matching functions CIEXYZ; and $\tau(\lambda)$ is the spectral function of transparent coefficients in the visible region.

The transparent coefficient is defined on known relationships:

$$\tau\ (\lambda) = 10^{-c * k\ (\lambda)} \tag{3}$$

k (λ)—absorption coefficient.

$$k\ (\lambda) = D/cL \tag{4}$$

where D is the optical density of solutions, c is the concentration, and L is the thickness of absorption layer.

For more hydrocarbon systems, the solutions were prepared with concentration near 0.002 g/l; similarly, as in this case, we got the result with minimal errors.

The relations for X, Y, Z are presented in matrix form [12]:

$$\begin{pmatrix} X \\ Y \\ Z \end{pmatrix} = \begin{pmatrix} E(\lambda_1)\overline{x}(\lambda_1) & E(\lambda_2)\overline{x}(\lambda_2) & \dots & E(\lambda_i)\overline{x}(\lambda_i) \\ E(\lambda_1)\overline{y}(\lambda_1) & E(\lambda_2)\overline{y}(\lambda_2) & \dots & E(\lambda_i)\overline{y}(\lambda_i) \\ E(\lambda_1)\overline{z}(\lambda_1) & E(\lambda_2)\overline{z}(\lambda_2) & \dots & E(\lambda_i)\overline{z}(\lambda_i) \end{pmatrix} \begin{pmatrix} \tau(\lambda_1) \\ \tau(\lambda_2) \\ \dots \\ \tau(\lambda_{i-1}) \\ \tau(\lambda_i) \end{pmatrix} \tag{5}$$

$$\mathbf{\Phi}_{XYZ} = \mathbf{E}_{XYZ}{}^{*}\,\mathbf{T} \tag{6}$$

where $\mathbf{\Phi}_{XYZ}$ is the column vector of color coordinates for objects of investigation in the system XYZ, $\mathbf{E}_{XYZ}$ is the product matrix of spectral power distribution for standard source and three color-matching functions, and **T** is the column vector for coefficient of transparency.

The chromaticity coordinates were calculated on formulas in the system CIE (12).

$$x = \frac{X}{X+Y+Z},\ y = \frac{Y}{X+Y+Z},\ z = \frac{Z}{X+Y+Z} \tag{7}$$

where x, y, and z are chromaticity coordinates.

In **Table 1** the defined CCs of multicomponent petrochemical systems are given [3, 4]. As it can be seen from the results of the calculations, CCs at the identical radiation source are close among themselves despite their different nature. Obviously, the reason of similarity of color properties is the similarity of the absorption spectra of the systems researched. Also the research has shown that multicomponent petrochemical systems do not have color isomerism, i.e., their CCs change depending on the radiation sources.

The coefficients B_0 and B_1 Eq. (1) have been calculated by the method of least squares. As the criterion of adequacy, the correlation coefficient R and the mean-square deviation have been taken. Some results of the calculations are given in **Table 2**. The received results show that for all the researched petrochemical systems, there is correlation dependence PCP from CCs [2–4].

In many processes, it is necessary to take express control of the PCPs. Therefore, the dynamic form of Eq. (8) has been investigated in the author's very last investigation of more than 300 of multicomponent hydrocarbon systems:

$$\Delta Z_k = b \cdot \Delta q_k \tag{8}$$

Hydrocarbon systems	CIE standard source	Luminosity	Coordinates of chromaticity		
			x	y	z
Separator oils of the Russian Federation (Bashkortostan, West Siberia, Tatarstan)	A	20.56–70.59	0.51–0.63	0.36–0.42	0.00–0.08
	B	17.24–66.60	0.39–0.51	0.42–0.49	0.00–0.19
	C	15.73–65.91	0.38–0.54	0.38–0.47	0.01–0.24
	D_{65}	17.17–67.42	0.37–0.51	0.39–0.50	0.01–0.23
Blown, residual, road, and structural petroleum	A	9.80–64.97	0.51–0.67	0.32–0.42	0.00–0.07
	B	7.76–61.44	0.40–0.55	0.44–0.46	0.01–0.16
	C	6.60–60.88	0.38–0.61	0.38–0.43	0.02–0.22
	D_{65}	7.48–62.79	0.38–0.57	0.41–0.45	0.01–0.21
Organic fractions of oligomers	A	68.23–87.09	0.50–0.52	0.39–0.40	0.08–0.11
	B	64.75–84.41	0.36–0.38	0.39–0.41	0.20–0.25
	C	64.40–84.71	0.34–0.37	0.35–0.37	0.26–0.32
	D_{65}	65.66–85.51	0.34–0.37	0.36–0.39	0.25–0.30
Residual high-boiling hydrocarbonic fraction of vacuum oil refining	A	15.26–86.19	0.51–0.66	0.33–0.40	0.00–0.09
	B	12.15–82.06	0.38–0.57	0.42–0.47	0.01–0.20
	C	10.41–81.90	0.36–0.62	0.36–0.44	0.01–0.26
	D_{65}	11.80–83.30	0.36–0.58	0.39–0.46	0.01–0.25
Hydrocarbonic fractions with average boiling temperature T_{boil} 180–360°C	A	21.20–99.83	0.48–0.63	0.36–0.41	0.01–0.14
	B	17.70–98.97	0.33–0.52	0.36–0.48	0.03–0.30
	C	15.98–99.96	0.31–0.55	0.32–0.45	0.04–0.37
	D_{65}	17.57–99.90	0.31–0.52	0.33–0.47	0.04–0.36
Asphaltenes and tars	A	0.10–98.67	0.29–0.86	0.14–0.41	0.00–0.33
	B	0.09–97.66	0.04–0.62	0.32–0.64	0.00–0.32
	C	0.08–98.67	0.04–0.71	0.24–0.59	0.00–0.37

Table 1.
Range of color characteristics of multicomponent hydrocarbon systems [3, 4].

where ΔZ is the change of the physicochemical property and Δq is the change of CCs. Eq. (8) means that a change of properties is proportional to the change of color for any colored substances.

The received results show that for all the researched petrochemical systems, there is correlation dependence PCP from CCs. The correlation coefficient R and the standard deviation were used as the criterion of adequacy. Some results of calculations are given in **Table 2**. Properties such as relative density (ρ); number-average molecular weight (M in Dalton); Conradson carbon residue (g in weight.%); activation energy for viscous flow (Ea in kJ/mol). The results show that for all studied petrochemical systems, there is a clear dependence of PCP on CCs [2–4]. These correlations allow the determination of PCP substances using CCs. Such dependencies are necessary for quality control of oil distillates and oil products. In addition, there is an opportunity for remote control methods of environmental pollution by oil and oil products.

For example, it is possible to determine in a few minutes such properties of formation oils as molecular mass, viscosity, density, the index of thermal stability, the index of reactivity of fractions in coking, thermal cracking processes, etc.

Multicomponent hydrocarbon system	PCP	CC	Coefficients of Eq. (1)		Correlation coefficient	Variation coefficient (%)	Fisher's ratio test for sample volume F
			B_0	B_1			
Raw oils	*p*	y_D	0.793	0.349	0.98	0.05	887.80
		g_A	0.758	0.310	0.98	0.05	889.46
	M	X_A	846.429	−4.563	0.99	0.48	1931.49
		R_A	897.646	−2.683	0.99	0.46	2084.20
	g	Y_B	17.063	−0.150	0.96	3.23	453.22
		R_C	17.984	−0.098	0.96	3.20	460.43
	Ea	Y_B	37.701	−0.376	0.97	5.17	536.05
		G_A	37.463	−0.140	0.97	5.16	536.51
Petroleum residues	*p*	x_C	0.757	0.462	0.99	0.27	577.35
		r_C	0.874	0.240	0.99	0.20	1019.05
	M	Y_B	877.611	−6.183	0.95	4.50	146.34
		r_A	−121.96	1157.340	0.96	4.31	161.29
	g	x_D	−34.205	106.697	0.98	6.80	341.48
		r_C	−6.530	53.960	0.98	5.92	455.11
	Ea	x_D	−76.698	228.968	0.98	8.31	308.21
		r_B	−22.755	110.594	0.98	7.53	378.34
Bitumens and bituminous materials	*p*	x_D	0.612	0.676	0.98	0.26	260.20
		r_C	0.876	0.209	0.99	0.22	384.50
	M	X_A	−11.918	1308.245	0.99	1.61	797.32
		R_A	1341.792	−6.249	0.99	1.47	958.95
	g	y_A	−570.815	255.283	0.98	3.62	252.80
		g_A	241.685	−379.399	0.98	3.64	249.38
	Ea	Y_A	−0.878	68.000	0.98	3.32	294.90
		G_A	64.355	−0.341	0.98	3.36	287.60

*ρ = *relative density;* M = *number-average molecular weight, moles;* g = *Conradson carbon residue, wt.%;* Ea = *activation energy for viscous flow, kJ/mol.*

Table 2.
Coefficients of Eq. (10) for physicochemical property estimation of oils and petroleum residues in colorimetric systems XYZ and RGB [3–10].

3. Introduction of electronic phenomenological spectroscopy

The method of electronic phenomenological spectroscopy (EPS) was first proposed by Mikhail Dolomatov [2, 3]. In recent years, this science direction has been intensively developed by the Dolomatov group at the Oil Technical State University and Bashkir State University (Ufa) in Russia. There are the following approaches and physical phenomena in the basis of EPS:

Unlike conventional spectroscopic methods, the EPS studies substances as a comprehensive quantum quasicontinuum without separating the spectrum of the substance into characteristic spectral bands by certain resonance frequencies or wavelengths of individual functional groups or components. The spectrum is studied as a single system (broadband signal) from a set of electronic states. Therefore, at this system integral level, there are new physical effects, not previously known. For example, the effects of the relationship of integral optical characteristics with different macroscopic and quantum properties of the substance as a whole by quantum quasicontinuum "spectrum-properties" and "color-properties" are observed. Qualitatively new physical phenomena appear when considering systems interacting with radiation in a wide optical spectrum. According to these laws, changes in the physical and chemical properties of substances cause a change in the integral characteristics of absorbed, reflected, or emitted radiation in the ultraviolet (UV), visible, and near-infrared (IR) regions of the electromagnetic spectrum. This allows the use of EPS methods for the study of individual and complex multicomponent substances.

For example, there may be a relationship between the integral force of the oscillator and some physical and chemical properties Z:

$$\Delta Z = \mathrm{c} \cdot \Delta \Theta \tag{9}$$

Here θ is the integral absorption—integral oscillator force (IOF), which has a simple physical explanation, namely, the area under the radiation absorption curve for the visible and UV regions of the spectrum, l nm mole^{-1} cm^{-1};

c is the constant depending on the method of measuring the spectrum, the nature of the substance, and individual for each property.

Therefore, it can be assumed that there is a relationship between any integral optical characteristic of a wide-spectrum signal (**Figure 4**) and properties having the form.

$$\Delta Z = \mathrm{c} \cdot \Delta P \tag{10}$$

Here, P is the integral spectral parameter, for example, integrated oscillator power, color characteristics, integral autocorrelation function, or relative imperial parameter and others into **Figure 1**.

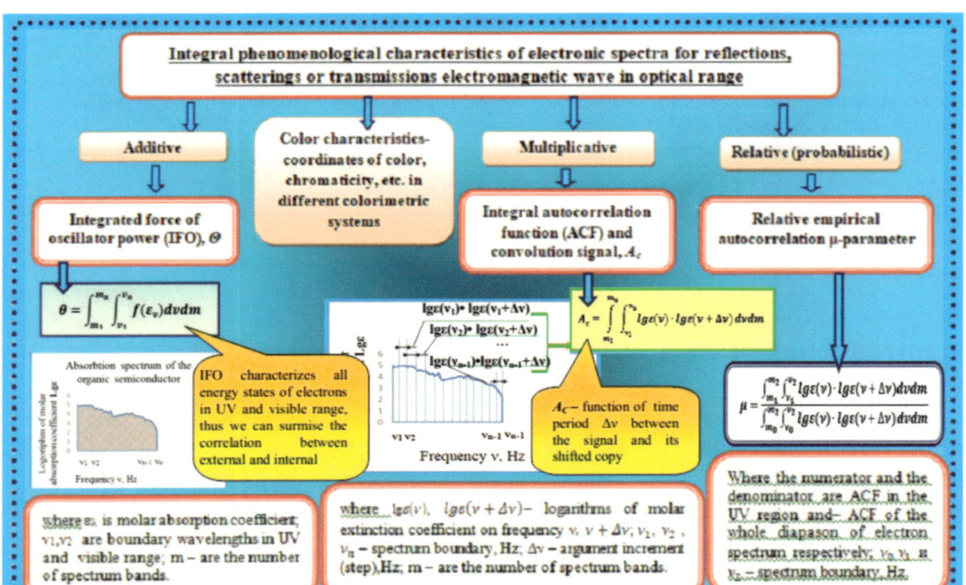

Figure 1.
Integral phenomenological characteristics of electronic spectra.

Obviously, a special case of Eq. (10) is the effect of "color-properties" (1) (we found with the coauthors O. Kydyrgychova, L. Dolomatova and V. Kartasheva in 1999 [5]). Phenomenological spectroscopy methods have been developed for identification and simultaneous determination of a set of different physical and chemical properties of natural and technical multicomponent organic systems, as well as properties of individual substances. For example, in a few minutes, it is possible to determine such properties of formation oils as the average molecular weight, viscosity, density, thermal stability index, index of reactivity of fractions in the processes of coking and thermal cracking, etc. EPS methods were adopted in the oil and petrochemical industry [2–4], environmental monitoring [3], biophysics and medicine [24, 25], nanotechnology and molecular electronics [15–18], and space exploration.

For science and technology, of interest are laws of the relationship of the integral characteristics of the spectrum and the electronic properties of matter. The knowledge of the electron structure of the molecular substances and materials has the fundamental importance for solving real problems in many fields of science and technology (physic of solid state, chemistry, electronics, electrical engineering). Despite progress in the experimental and quantum methods in some cases, there are significant discrepancies between the predicted values and experimental results of electron structure determination of complex materials and compounds. Many compounds and some materials for nanotechnology are characterized by complex structure and chemical and phase instabilities. Therefore, it is necessary to create new methods for assessing electronic structures, for example, ionization potentials, affinities to electron, and some other properties.

Hence the difficulty of determining the first ionization potentials (IP), the affinity of electrons (EA) and other characteristics of the energies of electronic states for such systems.

As known, ionization energy is the energy required to remove an electron from an atom or molecule. The unit of measurement of this physical quantity is the amount of energy required to remove one electron from one atom or molecule, expressed in electronic volts. The ionization potential (IP) is the electrical potential at which an electron leaves an atom or molecule, overcoming the forces of attraction. This process forms a positive ion [27, 28].

$$\mathbf{X - e^- \rightarrow X^+ + energy} \tag{11}$$

If during the electronic transition the geometry of the molecule changes minimally, it is said about the vertical IP. Next, we will consider the vertical potential only. According to the theorem of Koopmans, the first vertical ionization energy of a molecular system is equal to the negative of the orbital energy of the highest occupied molecular orbital (HOMO).

The electron affinity (EA) of an atom or molecule is defined as the amount of energy released or spent when an electron is added to a neutral atom or molecule with the formation of a negative ion [28].

$$\mathbf{X + e^- \rightarrow X^- + energy} \tag{12}$$

In the chemistry IP and EA are the characteristics for ability of molecules to donor-acceptor properties [27]. These physical values may be used for the determination of the indexes of reactivity of molecules (a characteristic of its chemical activity).

In previous works [2, 14, 20, 21], we established new physical effects between spectral densities (integral absorption, reflection, and transmission characteristics) with IP and EA. We propose to use these effects in determining the energies of electronic states. Methods for determination of IP and EA for molecules and organic semiconductors have been developed. We propose to use these effects in determining the energies of electronic states.

The IP and the EA of materials were estimated from the empirical dependencies linking these characteristics with the integral parameter of UV and/or vis spectrum:

$$\boldsymbol{E = \alpha_0 + \alpha_1 P} \tag{13}$$

where E is effective ionization potential or effective electron affinity, eV; α1 and α2 are empirically determined coefficients, and P is the integral spectral parameter. For example, integrated oscillator force (IOF), color characteristics, integral auto-correlation function or relative empirical parameter, and others (**Figure 1**).

The first experiments in the detection of the phenomenon (2) were carried out in 1988–1992 together with the Dr. G. Mukaeva [14]. The dependence of IP and EA on the integral oscillator force (IOF) was established by the results of the study of about 200 optical spectra of atoms and organic molecules:

$$E = \alpha_1 + \alpha_2 \cdot \theta, \tag{14}$$

Integral spectral characteristic can be any physical value of general absorption or emission of electromagnetic radiation, such as integral oscillator force (IOF):

$$\theta = \int_\delta \int_\xi f(\xi) d\xi d\delta \tag{15}$$

where θ is a reflection of quantum continuum as the sum of different states of electron, for example, all vibration states and electron states of transition among different levels, $f(\xi)$ is spectral function of absorption or emission of radiation, and δ is a range of resonance wavelengths or frequencies.

Let us consider the method, which was proved in our previous works [14, 15]. The IP and EA are estimated according to empirical dependencies which link these characteristics with logarithmic integral index of absorption (1).

$$E = \alpha_1 + \alpha_2 \theta_{\text{lg}}, \tag{16}$$

Here E is the ionization potential or an electron affinity, eV; α_1 and α_2 are empirically determined coefficients, eV and eV $\cdot$nm^{-1}, respectively.

$$\int_{\lambda_1}^{\lambda_2} \lg \varepsilon_\lambda d\lambda = \theta_{\text{lg}} \tag{17}$$

where $\varepsilon(\lambda)$ is the molar extinction coefficient, l mol^{-1} cm^{-1}; θ_{1g} is the integral logarithmic index of absorption (logarithmic IOS), ·nm; λ_1 and λ_2 are borders of the spectrum in UV and (or) visible region, nm; and λ_1 and λ_2 are the borders of wavelength of the spectrum in UV and (or) visible region.

Table 3 shows the corresponding coefficients for the dependencies (16) in different classes of organic molecules.

Breakthrough research in this area was done in collaboration with Dr. D. Shulyakovskii, Dr. E. Kovaleva, Dr. G. Yarmuhamedova, N. Paimurzina, and K. Latypov [20, 21]. We established the following regularities, which connected the integral parameters of the spectrum with IP and EA (18)–(21).

Dependence		$E = \alpha_1 + \alpha_2 \cdot \theta_{lg}$				
Homologous series	**IP or EA**	**Coefficient of correlation equations**		**Statistic characteristics**		
		α_1, eV	α_2, 10^{-7} eV nm^{-1}	**Correlation coefficient**	**Mean-square deviation, eV**	**Variation coefficient, %**
Polycyclic aromatic compounds	IP	8.074	−0.0010256	0.76	0.22	3.07
Polycyclic aromatic compounds	EA	0.290	0.00064502	0.71	0.16	2.22
Nitrogen-containing compounds [35]	IP	10.11	−0.00250000	0.88	0.26	2.46
Oxygen-containing compounds [35]	IP	11.03	−0.00347000	0.82	0.32	2.54

Table 3.
Coefficients of dependence (16) for homologous series.

$$IP = \gamma_1 + \gamma_2 \cdot A_{Cv}, \tag{18}$$

$$EA = \chi_1 + \chi_2 \cdot A_{Cv}, \tag{19}$$

$$IP = \varphi_1 + \varphi_2 \cdot \mu, \tag{20}$$

$$EA = \eta_1 + \eta_2 \cdot \mu, \tag{21}$$

IP is the effective ionization potential; EA is the effective electron affinity; A_{cv} is integral autocorrelation function of the electron spectrum (IAFS) (23); μ is the relative empirical autocorrelation parameter (μ, parameter) (24); ε (v) is the density distribution function of the radiation absorption; v is the spectral frequency; γ_0, γ_1, χ_1, χ_2, η_1, η_2, φ_0, and φ_1 are empirically defined coefficients (**Table 4**); and m is the number of spectrum bands.

In the calculation of integral parameters using the autocorrelation function of the signal, we have used the techniques adopted in statistical physics and spectroscopy [29]. We presented the energy spectrum of the molecule in the form of the integral of the autocorrelation function (IACF), frequency-dependent transitions. The integral autocorrelation function (ACF) is defined by the following formula:

Group of organic semiconductor	**Constants by (18) and (19)eV**		**Constants by (18) and (19), 10^{-17} eV s**		**Determination coefficient, R^2**	
	γ_1	χ_1	γ_2	χ_2	**IP**	**EA**
Complex oxy-compounds	9.35	0.08	−1.96	1.24	0.90	0.88
Ketones and aldehydes	10.65	−0.02	−2.98	1.76	0.85	0.81
Constants by (20) and (21), eV					Determination coefficient, R^2	
Polycyclic aromatic hydrocarbons	φ_1	φ_2	η_1	η_2	**IP**	**EA**
	5.43	1.68	1.88	−1.36	0.88	0.87

Table 4.
Constants and determination coefficients for dependencies (14–16).

$$A(\Delta\omega) = \int_{\omega_0}^{\omega_2} f(\omega) f(\omega + \Delta\omega) d\omega \qquad (22)$$

where $\omega = 2\pi v$, cyclical frequency, s^{-1}.

In [20] we proposed numerical parameter from IACP in the optical spectra was determined with the logarithmic function. The parameters of the ACF are because numbers are calculated using definite integral.

$$A_\nu = \int_{m_1}^{m_2} \int_{\nu 1}^{\nu 2} \varepsilon(\nu)\varepsilon(\nu + \Delta\nu) d\nu \mathrm{dm} \qquad (23)$$

where $\lg\varepsilon(\nu)$ and $lg\varepsilon(\nu + \Delta\nu)$ are logarithms of molar extinction coefficient on frequency ν, $\nu + \Delta\nu$; ν_1, ν_2, and ν_n are the spectrum boundary, Hz; $\Delta\nu$ is the argument increment (step), Hz; and m is the number of spectrum bands.

$$\mu = \frac{\int_{m_1}^{m_2} \int_{\nu_1}^{\nu_2} lg\varepsilon(\nu) \cdot lg\varepsilon(\nu + \Delta\nu) dv dm}{\int_{m_0}^{m_2} \int_{\nu_0}^{\nu_2} lg\varepsilon(\nu) \cdot lg\varepsilon(\nu + \Delta\nu) dv dm} \qquad (24)$$

where the numerator of fraction is the integral autocorrelation function (IACF) in the UV spectral region; the denominator is IACF in the UV-vis spectral region; ν, $\nu + \Delta\nu$; ν_1, ν_2, *and* ν_n are the spectrum boundary 10^{14} Hz; Δv is small increment of the argument (the analysis step of 1.5×10^{16} Hz); lgε(v) and lgε(v + Δv) are molar absorption coefficients at certain frequencies; and m is the number of spectrum bands.

The dependencies of IP and EA on the μ-factor for polycyclic aromatic hydrocarbons (PAH) of various classes (**Figures 2** and **3**) are established [20]. In addition, the dependencies of IP and EA on IACP for oxygen-containing compounds (alcohols, aldehydes, ketones) (**Figures 4** and **5**) are established [21].

IP and EA of organic molecules and PAH of different origin are presented in **Tables 5** and **6**.

EA and IP of organic molecules and semiconductors of different origins are presented in **Tables 2** and **3**.

Thus it is established that IP and EA of PAH, calculated by RHF 6-31G** and DFT methods, have the IACF dependence and μ-factor. These dependencies allow

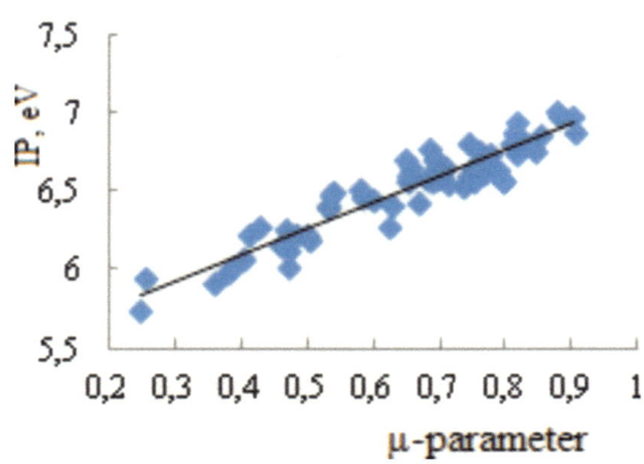

Figure 2.
Relationship of IP with the relative empirical autocorrelation parameter μ for PAH.

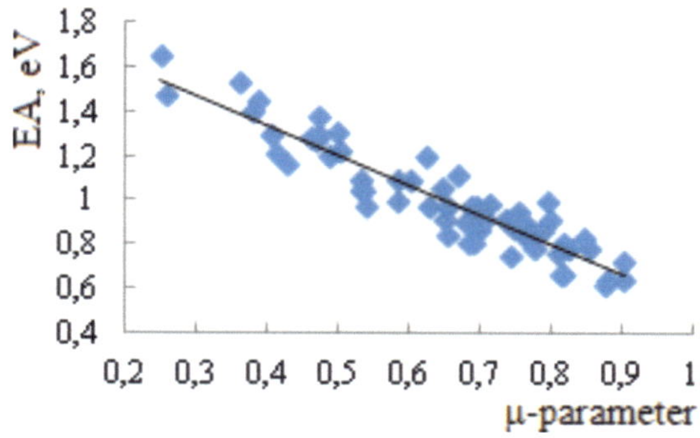

Figure 3.
Relationship of EA with the relative empirical autocorrelation parameter μ for PAH.

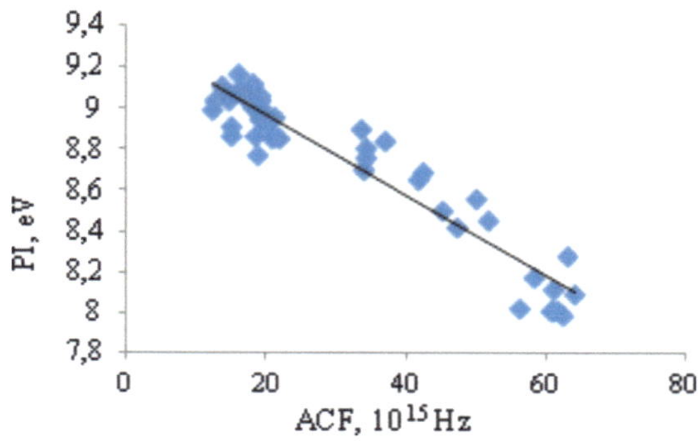

Figure 4.
Relationship of IP from IACF of organic oxygen groups containing molecules.

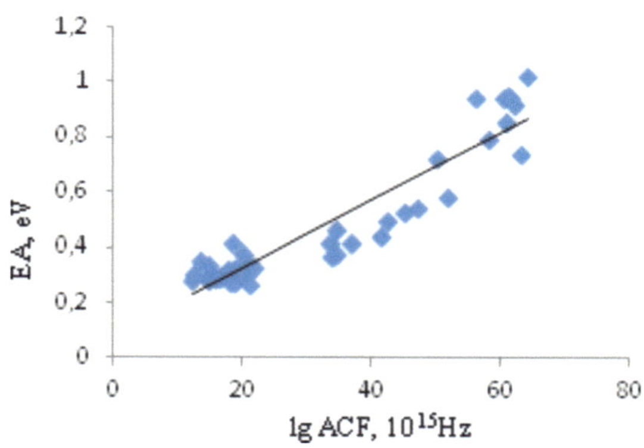

Figure 5.
Relationship of EA from IACF of organic oxygen groups containing molecules.

simplification of the estimations of IP and EA of organic molecules and PAH of different origin (**Tables 5** and **6**).

Thus, new methods for determining characteristics of electronic structure of different molecules and organic semiconductors are developed.

Subsequently, dependence (14) was confirmed by the study of IP and EA for various classes of sulfur and nitrogen organic compounds, organic dyes, amino acids, and biological fluids [24].

In studies [2–4] for very complex multicomponent systems, the problem of determining the electronic structure and, consequently, chemical activity was solved.

Molecules	μ-parameter	IP method DFT, eV	IP Eq. (4), eV	EA method DFT, eV	EA Eq. (5), eV
Hexahelicene	0.900	6.97	6.94	0.64	0.66
1.2,3.4,7.8-Tribenztetracene	0.820	6.82	6.81	0.78	0.77
Heptaphene	0.745	6.60	6.68	0.88	0.87
Pentacene	0.404	6.07	6.10	1.30	1.33
1,2-Benzpentacene	0.503	6.18	6.27	1.23	1.20
1,2-3,4-8,9-10,11-Tetrabenzpentacene	0.600	6.44	6.43	1.10	1.07
Naphtho-(2′.3′:3.4)-pyrene	0.647	6.70	6.52	1.06	1.00
3,4-Benznaphtho(2″,3″:8,9)-pyrene	0.472	6.12	6.71	1.30	0.84
3,4-Benznaphtho(2″,3″:9,10)-pyrene	0.531	6.41	6.20	1.05	1.26
1,14-4,5-Dibenzpentacene	0.765	6.66	6.69	0.86	0.86
1,2-Benzphenanthrene-(9′,10′:6,7)-pyrene	0.775	6.73	6.61	0.78	0.92
1,16-4,5-Dibenzhexacene	0.624	6.28	6.58	1.20	0.95
1,2-11,12-Dibenzperylene	0.378	5.98	6.06	1.41	1.37
1,12-2,3-Dibenzperylene	0.807	6.81	6.78	0.76	0.79
1,2-5,6-Dibenzcoronene	0.760	6.76	6.70	0.84	0.85

Table 5.
Calculated values of IP and EA.

Molecules	ACF, 10^{15} Hz	IP method HF, eV	IP Eq. (2), eV	EA method HF, eV	EA Eq. (3), eV
1-Phenylacetylbutadiene	55.46	9.01	9.00	0.99	0.96
2-Furylpolyenoic acids C_4H_3O—$(CH=CH)_2COOH$	61.49	8.92	8.95	1.13	1.06
Polyenoic acid	64.39	8.77	8.87	1.11	1.10
9-Oxoacridine	41.32	8.38	8.69	1.15	1.01

Table 6.
Calculated values of IP and EA.

The characteristics of the chemical activity can be determined from the electron absorption spectra simplification. The authors introduced new values: effective IP and effective electron affinity [30]. The effective IP and EA are the averaged potentials of ionization and the electron affinity of the radiation-absorbing components.

They allow to estimate the electron states of multicomponent and high-molecular substances, such as heavy residual resins of oil processing, high-molecular mixtures, and others.

Determining the electronic structure of materials and nanomaterials is an important problem of molecular electronics. For this, EPS was used. This application of EPS to determine the electronic structure of high-molecular compounds of petroleum (petroleum asphaltenes) was proposed in our previous works (Dolomatov et al.) [2–4, 30].

Asphaltenes	EIP, eV	EEA, eV	Band gap energy, eV	Quasi Fermi level, eV
Asphaltenes of Radevski oil	5.70	1.85	3.85	1.92
Asphaltenes of Surgut oil	5.20–5.70	2.10–2.50	3.10–3.20	1.55–1.60
Asphaltenes of distillate fraction	4.37–5.27	2.44–2.50	1.93–2.77	0.96–1.38
Hydrogenation asphaltenes of West Siberia oil	6.41	2.66	3.75	1.85
Asphaltenes and resins of Surgut oil	5.34	1.82	3.52	1.76
Asphaltenes of Kushkul oil	5.2	1.90	3.30	1.65

Table 7.
The characteristics of the electronic structure of asphaltenes by EPS method.

The asphaltenes are complex substances that can be found in crude oil, bitumen, and high-boiling hydrocarbon distillates. The asphaltenes are composed mainly of polyaromatic and heterocyclic compounds with traces of vanadium and nickel, which are in porphyrin structures. The electronic structure of asphaltenes has not been researched enough. The aim of research was to define the electronic structure of various asphaltenes. We have used the EPS methods. Some of the results are shown in **Table 7**.

Thus for asphaltenes, IP is in the interval from 4.37 up to 6.41 eV and EA differs from 1.82 to 2.66 eV. The size of energy band gap from 1.93 to 3.85 eV indicates that oil asphaltenes belong to amorphous, compensated, wideband semiconductors. The experiments for band gap estimation of the asphaltene molecules were confirmed by electronic structure computing with ab initio methods. The main deduction from this research is that oil asphaltenes can be used as organic semiconductors.

4. Effects of dependence of ionization potentials and electron affinity with color characteristics

The research [7] (co-author Dr. Shulyakovskaya D. and Dr. Yarmuhametova G.) established the phenomenon of the relationship between the energy of the molecular orbital, which characterizes the IP and EA, and color properties.

$$E = \beta_1 + \beta_2 \cdot q, \qquad (25)$$

where E is energy of the boundary molecular orbital (IP or EA), eV; α_1 and α_2 are empirically determined coefficients eV; q is one of the color characteristics (CCs) for standard light source A, B, C, or D; and CCs can be represented in one of the international color measurement systems (e.g., color coordinates or chromaticity coordinates in XYZ or RGB systems). The color coordinates of polycyclic aromatic hydrocarbons in the XYZ system are shown (**Figure 6**). These coordinates are calculated in the visible region of the transmission spectra of hydrocarbon solutions according to the formulas (2)–(7).

Several classes of compounds, including PAH, were studied by dependence (25). The corresponding coefficients for IP and EA are presented in **Tables 8** and **9**. As can be seen from the tables, the accuracy of the assessment of ionization potentials and electron affinity is satisfactory. Thus, the effect of the relationship between IP and EA on the color characteristics can be used to simultaneously measure these physical quantities.

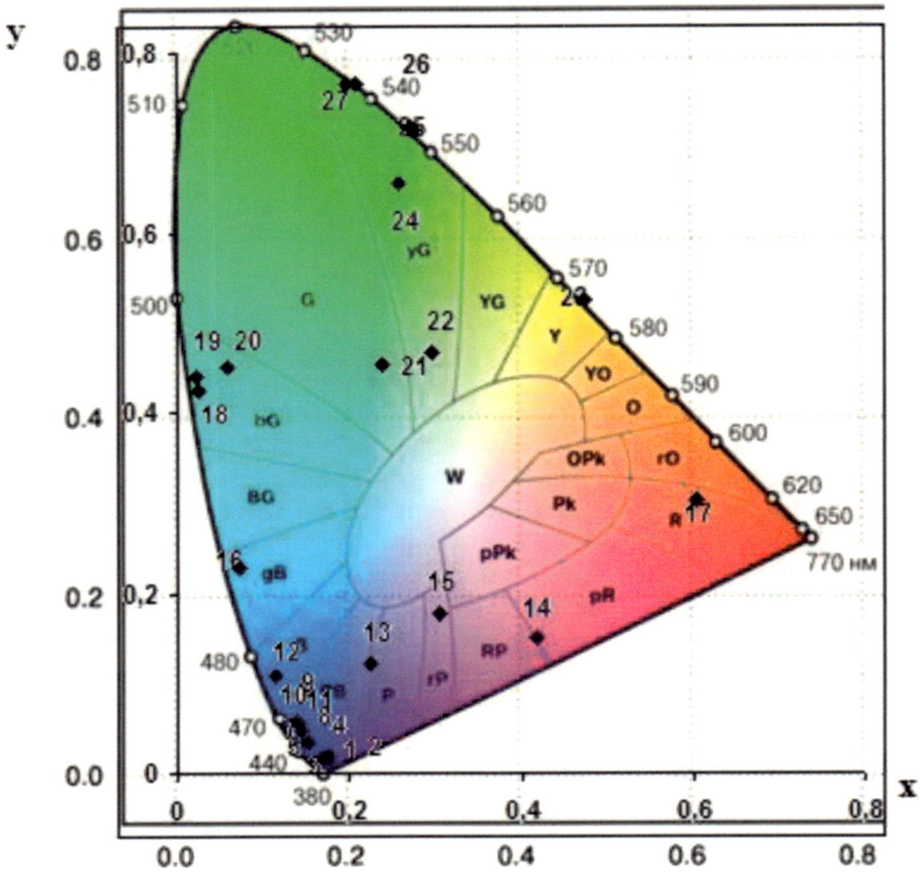

Figure 6.
Color characteristics (chromaticity coordinates x and y) of the individual aromatic oil components in XYZ colorimetric system: (1) perylene, (2) tetrabenzpentacene, (3) dibenzpyrene, (4) hexabenzcoronene, (5) 1,2-benzphenantrenopyrene, (6) 2,3-benzperylene, (7) dibenzpentacene, (8) phenantrenopyrene, (9) ovalen, (10)–(12) dibenzperylenes, (13) dibenzpyrene, (14) dinaphtpyrene, (15) tetrabenzheptacene, (16) benzanathtpyrene, (17) dibenzanthanthrene, (18) bisantene, (19) benzanathtpyrene, (20) benzbisantene, (21) dinathteptacene, (22) 1, 2-benzanaphtpyrene, (23) dibenzperylene, (24) dibenzanthanthrene, (25) benzperylene, (26) dinaphtpyrene, and (27) tetrabenzheptacen.

Organic semiconductor class	CCs	Coefficients for IP		Correlation coefficients	Variation coefficients (%)	Standard deviation (eV)
		A_1 (eV)	A_0 (eV)			
Semiconductors containing three and five linear annelated benzene rings and semiconductors of perylene series	Z_C	−0.0120	8.2188	0.90	3.10	0.23
	Z_D	−0.0129	8.2035	0.89	3.13	0.23
	B_C	−0.0023	8.2256	0.89	3.17	0.24
	B_D	−0.0024	8.2110	0.89	3.20	0.24
Semiconductor of bisantene series and anthanthrene	z_C	4.7985	3.7638	0.94	4.44	0.31
	z_D	4.5947	3.9328	0.94	4.47	0.31
	b_B	4.2833	3.4563	0.94	4.71	0.32
	b_C	5.3597	2.4476	0.94	4.52	0.31
Semiconductors of pyrene series	x_C	−4.2636	7.8232	0.87	2.85	0.20
	x_D	−4.2503	7.8231	0.86	2.88	0.20

Organic semiconductor class	CCs	Coefficients for IP		Correlation coefficients	Variation coefficients (%)	Standard deviation (eV)
		A_1 (eV)	A_0 (eV)			
	R_A	−0.0110	7.2866	0.87	2.86	0.20
	R_B	−0.0148	7.3764	0.87	2.80	0.20
Heterocyclic semiconductors	y_C	−1.9421	7.7117	0.94	1.92	0.14
	y_D	−1.8822	7.7100	0.93	1.96	0.14
	g_A	−1.1612	7.5854	0.91	2.32	0.17
	g_B	−1.2637	7.5599	0.90	2.41	0.18

Table 8.
Coefficients of dependence (25) for IP by PAH [4, 7].

Organic semiconductor class	CCs	Coefficients (25) for EA		Correlation coefficients	Variation coefficients (%)	Standard deviation (eV)	Sample volume (pcs)
		B_1 (eV)	B_0 (eV)				
Semiconductors containing three and five linear annelated benzene rings and semiconductors of perylene series	Z_C	0.0049	0.6344	0.90	9.92	0.09	29
	Z_D	0.0053	0.6407	0.89	10.01	0.10	
	B_C	0.0009	0.6316	0.89	10.77	0.10	
	B_D	0.0010	0.6376	0.89	10.88	0.10	
Semiconductor of bisantene series and anthanthrene	z_C	−1.9716	2.4650	0.94	10.65	0.13	11
	z_D	−1.8879	2.3955	0.94	10.71	0.13	
	b_B	−1.7597	2.5912	0.94	11.29	0.13	
	b_C	−2.2019	3.0056	0.94	10.85	0.13	
Semiconductors of pyrene series	x_C	1.7519	0.7970	0.87	7.62	0.08	20
	x_D	1.7464	0.7970	0.86	7.68	0.08	
	R_A	0.0045	1.0175	0.87	7.64	0.08	
	R_B	0.0061	0.9806	0.87	7.49	0.08	
Heterocyclic semiconductors	y_C	0.7978	0.8430	0.94	5.71	0.06	15
	y_D	0.7732	0.8437	0.93	5.82	0.06	
	g_A	0.4769	0.8949	0.91	6.89	0.07	
	g_B	0.5190	0.9054	0.90	7.17	0.07	

Table 9.
Coefficients of dependence (25) for EA by PAH [4, 7].

The dependence of the IP on the chromatic coordinate-Z in the XYZ system for PAH based on three and five linear annular benzene rings and from the perilene series is shown in **Figure 7**.

The IP and EA values for various organic molecules obtained by the dependence (11) are confirmed by various modifications of quantum DFT and ab initio methods. In addition, the values of IP were estimated by photoelectron spectroscopy. The results are shown in **Tables 10** and **11**.

From the received results, it follows that the equation is distributed to substances with IP < 9.8 eV, i.e., it covers the majority of organic substances.

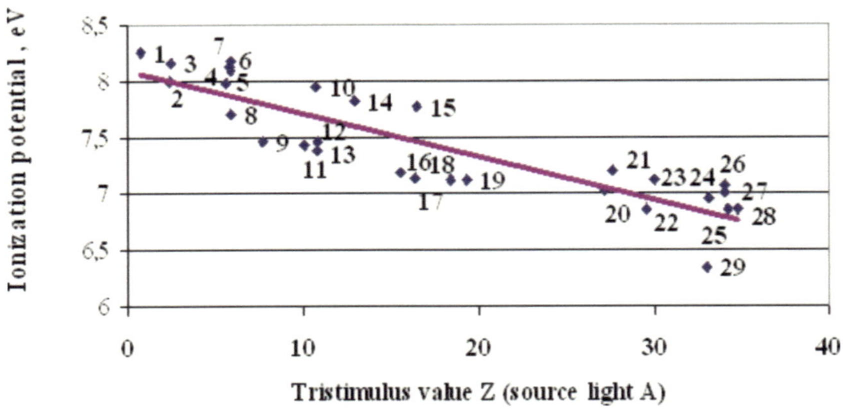

Figure 7.
The correlation of the first PI and the color characteristic for the compounds with three and five linear annulary benzene rings and from the perilene series. (1) 2,3-benzpizene, (2) 1,12-2,3-8,9-tribenzperylene, (3) 1,12-2,3-dibenzperylene; (4) anthracene [2′,1′:1,2] anthracene; (5) coronene; (6) 2,3–8, 9-dibenzpizene; (7) 3,4-benzpentaphene; (8) pentaphene; (9) perilene; (10) naphtha[2′, 3′:3, 4]pentaphene; (11) 1,12–o-phenylenperilene; (12) 1,2-benzcoronene; (13) 1,2-3,4-5,6-10,11-tetrabenzanthracene; (14) 2,3-8,9-dibenzpizene; (15) 1,2-7,8-dibenzcoronene; (16) 1,12–o-phenyl-2,3-10,11-dibenzperilene; (17) naphtha [2′, 3′:1, 2] coronene; (18) 2,3–10, 11-dibenzperylene; (19) 2,3-benzperylene; (20) 1,2-3,4-5,6-tribenzcoronene; (21) anthracene [2′, 1′,1, 2]tetraphene; (22) 1,2-benzperylene; (23) 1,2-10,11-dibenzperylene; (24) 1,2-3,4-8,9-10,11-tetrabenzpentazene; (25) 1,2-11,12-dibenzperylene; (26) 1,2-8,9-dibenzpentazene; (27) 1,2-benzpentazene; (28) pentazene; and (29) 1,2-7,8-dibenzpizene.

Organic semiconductor class	Semiconductor name	Electron affinity, eV		Abs. accuracy, *eV*	Rel. accuracy, %
		Regular methods	Acc. to CCs		
Semiconductors containing three and five linear annelated benzene rings and semiconductors of perylene series	Pentaphene	0.85	0.78	0.07	8.24
	Anthraceno [2′,1′:1,2]anthracene	0.73	0.77	0.04	5.75
	2,3-Benzpicene	0.62	0.70	0.08	12.81
	Anthraceno [2′,1′:1,2]tetraphene	1.05	1.12	0.06	6.05
	Pentacene	1.19	1.23	0.04	3.36
	1,2-Benzpentacene	1.13	1.22	0.09	7.71
	1,2-3,4-8,9-10,11-Tetrabenzpentacene	1.16	1.20	0.05	4.23
	1,2-Benzperylene	1.19	1.15	0.04	3.55
	1.2–10.11-Dibenzperylene	1.09	1.16	0.07	6.36
	1.2–11.12-Dibenzperylene	1.19	1.22	0.03	2.59
	1,12-2,3-Dibenzperylene	0.66	0.72	0.06	9.75
	1,12-2,3-8,9-Tribenzperylene	0.73	0.72	0.01	0.92
	1,2-3,4-5,6-Tribenzcoronene	1.12	1.11	0.01	1.13

Organic semiconductor class	Semiconductor name	Electron affinity, eV		Abs. accuracy, eV	Rel. accuracy, %
		Regular methods	Acc. to CCs		
Semiconductor of bisantene series and anthanthrene	Bisantene	1.69	1.71	0.02	1.33
	1,14-Benzbisantene	1.11	1.18	0.07	6.32
	3,4-11,12-Dibenzbisantene	0.88	0.93	0.05	5.44
	3,4–10.11-Dibenzbisantene	0.99	0.95	0.04	4.27
	1,2-3,4-8,9-10,11-Tetrabenzbisantene	0.89	0.97	0.08	9.03
	Anthanthrene	1.04	0.92	0.12	11.64
	1,2-7,8-Dibenzanthanthrene	0.95	0.95	0.00	0.02
Semiconductors of pyrene series	3,4–8.9-Dibenzpyrene	1.08	1.03	0.04	3.91
	3,4-9,10-Dibenzpyrene	1.01	1.02	0.02	1.59
	3,4-Benzanaft [2″,3″:8.9]pyrene	1.02	1.02	0.00	0.36
	3,4-Benzanaft [2″,3″:9,10]pyrene	0.97	1.03	0.06	5.93
	Dinaft[2′,3′:3,4]-[2″,3″:9,10]pyrene	1.00	1.01	0.01	0.97
	1,14-4,5-Dibenzpetacene	1.10	1.04	0.07	6.02
	5,6-15,16-Dibenzhexacene	1.06	1.04	0.03	2.65
	Naft[1′,7′:2,16] hexacene	1.40	1.40	0.00	0.16
	1,18-4,5-9,10-13,14-Tetrabenzheptacene	1.02	1.02	0.00	0.15
Heterocyclic semiconductors	9-Anthracentiol	0.91	0.93	0.02	2.38
	2,2′;5′,2”-Tertienil	0.83	0.87	0.05	5.46
	2-Tiapyranthion	1.26	1.30	0.04	3.22
	1,3-Ditiolene-2-thione	0.83	0.90	0.07	8.78
	4,5-Cyclohexeceno-1.3-ditiolene-2-thione	0.85	0.91	0.06	7.26
	4-Phenyl-1,3-ditiolene-2-thione	0.94	0.90	0.04	4.23
	Nafto[1,2-b]-1,3-ditiolene-2-thione	0.97	0.90	0.07	7.48
	4,5-Cyclopenteno-1.2-ditiolene-3-thione	1.18	1.12	0.06	5.12

Organic semiconductor class	Semiconductor name	Electron affinity, eV		Abs. accuracy, *eV*	Rel. accuracy, %
		Regular methods	Acc. to CCs		
	4,5-Cyclohexeceno-1,2-ditiolene-3-thione	1.12	1.12	0.00	0.21
	4,5-Cyclohepteno-1,2-ditiolene-3-thione	1.18	1.13	0.06	4.75
	Thiolane-3,4-dithion	1.01	1.00	0.01	1.47

Table 10.
Results of determining electron affinity of some organic semiconductors [7].

Organic semiconductor class	Semiconductor name	Ionization potential, eV		Abs. accuracy, eV	Rel. accuracy, %
		Regular methods	Acc. to CCs		
Semiconductors containing three and five linear annelated benzene rings and semiconductors of perylene series	Pentaphene	7.70	7.87	0.17	2.22
	Anthraceno [2′,1′:1,2]anthracene	7.99	7.88	0.10	1.26
	2,3-Benzpicene	8.26	8.07	0.19	2.34
	Anthraceno [2′,1′:1,2]tetraphene	7.20	7.04	0.15	2.14
	Pentacene	6.87	6.77	0.10	1.41
	1,2-Benzpentacene	7.01	6.80	0.21	3.01
	1,2-3,4-8,9-10,11-Tetrabenzpentacene	6.95	6.83	0.12	1.72
	1,2-Benzperylene	6.87	6.97	0.10	1.48
	1.2–10,11-Dibenzperylene	7.12	6.95	0.17	2.38
	1,2-11,12-Dibenzperylene	6.87	6.79	0.08	1.10
	1,12-2,3-Dibenzperylene	8.16	8.00	0.16	1.92
	1,12-2,3-8,9-Tribenzperylene	7.99	8.00	0.02	0.20
	1,2-3,4-5,6-Tribenzcoronene	7.03	7.06	0.03	0.42
Semiconductor of bisantene series and anthanthrene	Bisantene	5.66	5.60	0.05	0.96
	1,14-benzbisantene	7.06	6.89	0.17	2.42
	3,4-11,12-Dibenzbisantene	7.61	7.50	0.12	1.53
	3,4–10.11-Dibenzbisantene	7.35	7.46	0.10	1.40

Organic semiconductor class	Semiconductor name	Ionization potential, eV		Abs. accuracy, eV	Rel. accuracy, %
		Regular methods	Acc. to CCs		
	1,2-3,4-8,9-10,11-Tetrabenzbisantene	7.60	7.41	0.19	2.56
	Anthanthrene	7.24	7.53	0.29	4.06
	1,2-7,8-Dibenzanthanthrene	7.46	7.46	0.00	0.02
Semiconductors of pyrene series	3,4-8,9-Dibenzpyrene	7.14	7.25	0.10	1.44
Semiconductors of pyrene series	3,4-8,9-Dibenzpyrene	7.14	7.25	0.10	1.44
	3,4-9,10-Dibenzpyrene	7.32	7.28	0.04	0.54
	3,4-Benzanaft [2″,3″:8,9]pyrene	7.29	7.28	0.01	0.13
	3,4-Benzanaft [2″,3″:9,10]pyrene	7.40	7.25	0.14	1.90
	Dinaft[2′,3′:3,4]-[2″,3″:9,10]pyrene	7.33	7.31	0.02	0.32
	1,14-4,5-Dibenzpetacene	7.08	7.24	0.16	2.28
	5,6-15,16-Dibenzhexacene	7.17	7.24	0.07	0.98
	Naft[1′,7′:2,16] hexacene	6.35	6.36	0.01	0.10
	1,18-4,5-9,10-13,14-Tetrabenzheptacene	7.29	7.28	0.00	0.07
Heterocyclic semiconductors	9-Anthracentiol	7.54	7.49	0.05	0.71
	2,2′;5′,2”-Tertienil	7.75	7.64	0.11	1.40
	2-Tiapyranthion	6.69	6.60	0.10	1.46
	1,3-Ditiolene-2-thione	7.76	7.58	0.18	2.27
	4,5-Cyclohexeceno-1,3-ditiolene-2-thione	7.69	7.54	0.15	1.95
	4-Phenyl-1,3-ditiolene-2-thione	7.48	7.58	0.10	1.29

Table 11.
Results of determining of the first ionization potentials of some organic semiconductors [7].

Thus, it can be concluded that the effects (21)–(25) discovered by us allow us to estimate the energy levels of quantum systems with sufficient accuracy. This is important for the study of multi-electron systems in molecular electronics and nanotechnology and chemistry such as single molecules, atomic clusters, and high-molecular systems. From here it follows that electronic spectra and color characteristics can be applied to the definition of various characteristics of substances.

5. Normal and pathological color characteristics of human blood components

This cycle of works is described [22–26] and executed together with Dr. N. Kalashchenko and Dr. S. Dezortsev. The experiments were conducted at the Ufa Medical University and the Republican Clinic named after Kuvatov (Ufa, Russia).

Colorimetric studies of blood are actively used in medicine [31], criminal law [32], and the food industry [33, 34]. In medical practice, colorimetric methods are used to determine the hemoglobin concentration in the blood of a patient (the color index) [35]. Today a rather exact (±1%) cyanomethemoglobin photometric method is used everywhere, in which cyanomethemoglobin is determined at a wavelength of 540 nm after preparation of a working solution of the blood in Drabkin reagent. Various modifications of this method do not change its essential physical nature [6]. Furthermore, spectral analysis in the visible region has been used to determine oxyhemoglobin and other hemoglobin-containing compounds from the absorption spectra of blood and its solutions [36]. Despite this, the quantitative colorimetric characteristics of blood have not been studied before.

The aim of this work was to study the color characteristics of hemolyzed blood, plasma, and serum from donors in the visible range of the absorption spectra by standard CIE methods (International Commission on Illumination, 1964).

The basic color characteristics (lightness and chromaticity coordinates) determine the position of the color of the specimen in an arbitrary color space and are found by the CIE method [11, 12].

The familiar spectrophotometric method for color measurements involves measuring the spectral power distribution of the radiation followed by calculation of the color coordinates by multiplying the determined spectral power distribution function times the three color-matching functions and then integrating the products. For the spectral power distribution function of the source E(λ), the spectral transmittance function τ(λ), and x(λ), y(λ), and z(λ) (the color-matching functions) and the color coordinates X, Y, and Z are determined by integration over the wavelength range for visible radiation 380–760 nm. In practice, integration is replaced by summation over the interval dλ (from 5 to 10 nm), since the spectral functions under the integral sign are usually not easily integrated:

$$
\begin{aligned}
X &= \Delta\lambda \sum_{\lambda} E(\lambda)\tau(\lambda)\overline{x}(\lambda) \\
Y &= \Delta\lambda \sum_{\lambda} E(\lambda)\tau(\lambda)\overline{y}(\lambda), \\
Z &= \Delta\lambda \sum_{\lambda} E(\lambda)\tau(\lambda)\overline{z}(\lambda)
\end{aligned}
\tag{26}
$$

The spectral power distribution and the spectral transmittance curve are measured by separating light into a spectrum, such as in a spectrophotometer or monochromator. The color-matching curves are specified as tables of values of the specific coordinates in 10 nm steps. There are also tables of E(λ)x(λ) values for standard CIE light sources A, B, C, and D, characterizing the most typical natural (B, C, D) and artificial (A) illumination conditions.

The chromaticity coordinates are calculated using the formulas.

$$
x = \frac{X}{X+Y+Z}, \mathrm{y} = \frac{Y}{X+Y+Z}, z = \frac{Z}{X+Y+Z}, \mathrm{x+y+z=1} \tag{27}
$$

The coordinate Y characterizes the lightness (luminance) of the specimens.

The quantitative colorimetric characteristics of hemolyzed blood, plasma, and serum described by formulas (26) and (27) in the standard CIE method are connected with the transmittance or reflectance spectra and are integrated parameters determined over the entire visible region of the electromagnetic spectrum. So it is assumed that they carry information about the condition of the entire body. In our approach, blood and its components are considered as a single, indivisible light-absorbing system.

The experiment. The objects of investigation were solutions of hemolyzed blood and solutions of plasma and serum (prepared from that blood) of the same concentration from 100 male and female donors (in different blood groups and age groups) and from 95 patients who were assigned to three arbitrary groups: (I) 41 patients with purulent diseases (osteomyelitis, purulent fistulas, gonitis), (II) 41 resuscitated patients (acute myocardial infarction, acute cerebral circulatory collapse, chronic cardiac insufficiency), and (III) 13 patients with cirrhosis of the liver. We determined the color characteristics of the "average" donor (without separating the donors according to blood, sex, and age groups) and compared them with the analogous characteristics of patients from the different groups. The blood for the studies was drawn at a blood donation center and in clinical departments by standard procedures [37].

The spectra of the solutions of hemolyzed blood, plasma, and serum of concentration 2.5 vol.% (1.40) were taken in quartz cuvets with thickness of the working layer of liquid equal to 10 mm at room temperature, on an SF-2000 spectrophotometer in the range 200–1000 nm in 20 nm steps. As the solvent and the reference solution, we used distilled water for injection, which is optically neutral under the experimental conditions and is the natural physiological solvent in the human body. The hemolyzed blood was prepared using a standard heparin solution, and the plasma was prepared using the preservative Glyugitsir.

In addition to the averaged values of the color coordinates and the lightness value, we calculated the standard deviation, the confidence interval for significance level α = 95%, and the coefficient of variation.

Figure 8 shows the averaged spectra for the hemolyzed blood, plasma, and serum from the patients in all three examined groups compared with the corresponding averaged spectra of the donors. There are clear differences between the different groups of patients.

For the plasma and serum, over the entire studied region, the group spectra for the patients lie higher than the averaged spectra of the donors, and their positional order is consistent: the averaged spectrum for patients with purulent diseases lies above the averaged spectrum for the donors, the averaged spectrum for the

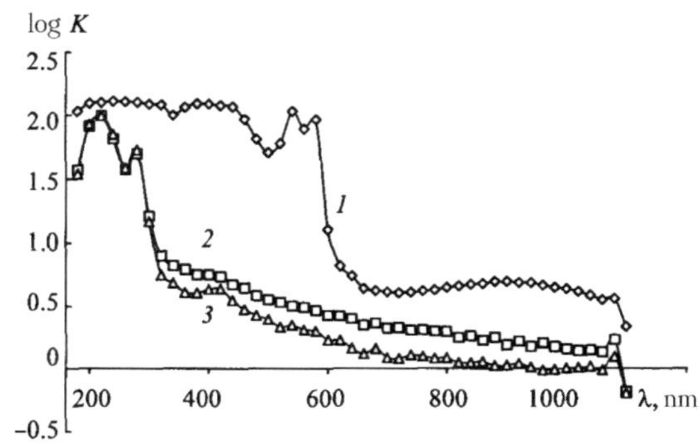

Figure 8.
Spectra of hemolyzed donor blood (1), plasma (2), and serum (3) in the UV and visible regions (averaged over 100 donors).

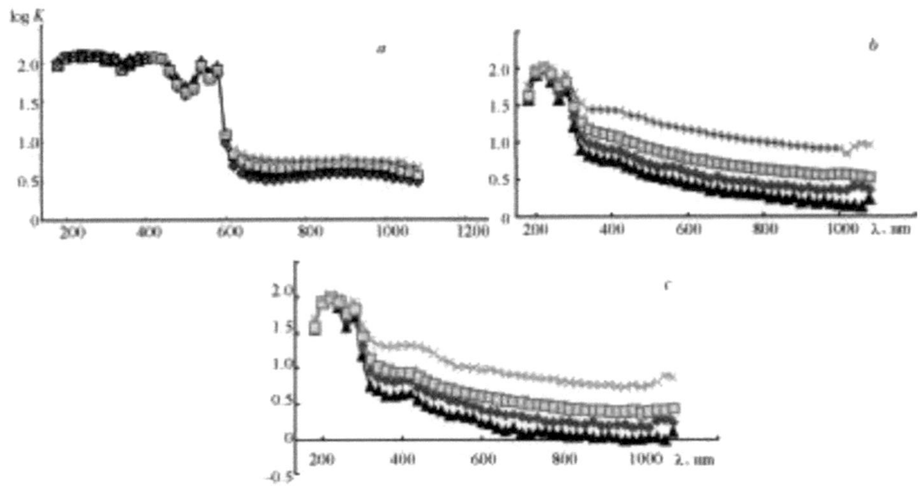

Figure 9.
Averaged spectra of hemolyzed blood (a), plasma (b), and serum (c) of examined groups of patients (▲ = donors, ¨ = I, ¤ = II, × = III) compared with averaged spectrum of hemolyzed blood, plasma, and serum, respectively, from donors [25].

resuscitation patients lies above that spectrum, and the averaged spectrum for patients with cirrhosis of the liver lies even higher. Probably such positioning of the spectra reflects the severity of the general condition of the patients, if we assume that cirrhosis is the most severe condition for the patients with the least likelihood of recovery. We do not observe such a dependence for the hemolyzed blood: the averaged spectrum for the patients with purulent diseases lies below the averaged spectrum for the donors (**Figure 9**).

Table 12 gives the averaged color coordinates and lightness for the donors and each group of examined patients, calculated for the solutions of blood, plasma, and serum as a single light-absorbing system according to the standard CIE method. For the donors, the chromaticity coordinate x varies from 0.320 ± 0.001 (for serum and plasma) to 0.630 ± 0.008 (for hemolyzed blood). The coefficients of variation in this case also decrease from 4.7 for blood down to 1.3 for serum. The chromaticity coordinate y for the donors has similar values: 0.320 ± 0.002 for plasma and serum and 0.340 ± 0.003 for blood. The coefficients of variation for y steadily decrease from 3.2 for blood down to 2.0 for serum. The parameter z for the donors is higher for serum and plasma (0.360 ± 0.003) than for blood (0.030 ± 0.005). The coefficient of variation for this parameter is maximum for blood (67.6) and minimum for serum (2.5). The lightness, as expected, has the maximum value (84.88–1.54) for serum and the minimum value (11.55–0.67) for hemolyzed blood. For plasma, this parameter is close to the value typical of serum.

In determining the color range (see **Figure 9a**) for the dilute solutions (1.40) of hemolyzed blood, plasma, and serum from the donors, the color range of blood falls within the red region of the spectrum; the range for plasma and serum falls within the yellow region with lower saturation, which supports the correctness of our experiments and calculations. The corresponding regions for the color range for the patients cover a larger area than for the donors (**Figure 9b**). In order to better visualize the results obtained, we calculated the color coordinates for the studied specimens with correction for concentration. All the points for the donors lie within the yellow-orange region with saturation of 30–50%, which corresponds to the visual observations.

The average values of the color coordinates for all the patient groups (see **Table 12**) are virtually no different from the averages for the donors except for

Parameter	Hemolyzed blood				Plasma				Serum			
	Donors	I	II	III	Donors	I	II	III	Donors	I	II	III
Chromaticity coordinate x												
Mean value	0.63	0.59	0.64	0.34	0.32	0.335	0.33	0.35	0.32	0.32	0.32	0.35
Standard deviation	0.029	0.041	0.032	0.048	0.005	0.009	0.013	0.02	0.004	0.009	0.016	0.019
Confidence interval, $\alpha = 0.95$	0.01	0.01	0.01	0.02	0.001	0.002	0.004	0.01	0.001	0.002	0.005	0.009
Coefficient of variation	4.7	6.9	5.46	8.11	1.6	2.65	4.13	5.82	1.3	2.9	4.93	5.5
Chromaticity coordinate y												
Mean value	0.34	0.35	0.35	0.36	0.32	0.33	0.34	0.36	0.32	0.33	0.34	0.37
Standard deviation	0.011	0.013	0.011	0.014	0.008	0.011	0.011	0.019	0.006	0.011	0.012	0.019
Confidence interval, $\alpha = 0.95$	0.003	0.003	0.003	0.007	0.002	0.003	0.004	0.009	0.002	0.003	0.004	0.009
Coefficient of variation	3.2	3.67	3.07	3.98	2.6	3.36	3.37	5.27	2.0	3.55	3.61	5.22
Chromaticity coordinate z												
Mean value	0.03	0.005	0.05	0.06	0.36	0.34	0.33	0.29	0.36	0.35	0.34	0.29
Standard deviation	0.02	0.028	0.024	0.034	0.012	0.017	0.024	0.039	0.009	0.019	0.026	0.038
Confidence interval, $\alpha = 0.95$	0.005	0.007	0.008	0.017	0.003	0.004	0.008	0.019	0.003	0.005	0.008	0.019
Coefficient of variation	67.6	53.77	53.16	61.58	3.3	4.98	7.29	13.46	2.5	5.53	7.76	13.26
Lightness L, %												
Mean value	11.55	14.82	12.99	13.46	78.94	67.82	63.04	36.66	84.88	79.34	72.42	48.08
Standard deviation	2.46	3.5	2.72	5.8	8.17	12.42	14.79	19.65	5.58	10.21	16.46	22.77
Confidence interval, $\alpha = 0.95$	0.67	0.9	0.85	2.84	2.19	3.11	4.62	9.63	1.54	2.62	5.14	11.15
Coefficient of variation	21.3	23.86	21.38	43.07	10.3	18.11	23.92	53.6	6.6	13.03	23.2	47.35

Note: I, II, and III indicate the arbitrary groups of patients.

Table 12.
Integrated normal and pathological color characteristics of hemolyzed human blood, plasma, and serum [22–25].

Samples of blood plasma	Patients with liver cirrhosis		Healthy people	
	x	y	x	y
1	0.348	0.350	0.322	0.324
2	0.345	0.348	0.324	0.326
3	0.341	0.343	0.318	0.315
4	0.383	0.373	0.331	0.325
5	0.348	0.357	0.321	0.315
6	0.339	0.348	0.323	0.317
7	0.360	0.368	0.315	0.316
8	0.350	0.363	0.322	0.326
9	0.372	0.375	0.33	0.331
10	0.342	0.347	0.309	0.31
11	0.344	0.346	0.313	0.311
12	0.359	0.356	0.326	0.325
13	0.345	0.347	0.317	0.318
14	0.357	0.359	0.314	0.317
15	0.353	0.360	0.319	0.325

Table 13.
The chromaticity coordinates (x and y) of patients with liver cirrhosis and healthy people [25].

Statistical indicators	Patients with liver cirrhosis		Healthy people	
	x	y	x	y
Mean	0.352	0.356	0.320	0.320
Confidence interval	0.006	0.005	0.001	0.002
Dispersion, σ	0.02	0.02	0.005	0.008
Error of mean	0.003	0.002	0.001	0.001
Variation coefficient	5.62	4.80	1.6	2.6

Table 14.
The statistical characteristics of chromaticity coordinates for blood plasma of healthy individuals and patients with liver cirrhosis in the XYZ system [25].

patients with cirrhosis of the liver, for which the bilirubin blood concentration sharply increases. As a result of this, the plasma and serum take on a saturated orange color, which is reflected in the spectra and accordingly in their color characteristics. For the same reason, the lightness of the plasma and serum for these patients is much lower (by almost a factor of two) than for the rest. However, we do not observe sharp differences in the spectra of the hemolyzed blood.

The lightness of the plasma and serum depends on the condition of the patients. Thus the average values of this parameter for the plasma decrease as the severity of the diseases increases: from 78.94 for the donors up to 67.82 for septic patients, 63.04 for resuscitated patients, and 36.66 for patients with cirrhosis of the liver.

In the future, we performed new experiments in which the possibility of diagnosis of liver cirrhosis by color characteristics of blood was considered [23].

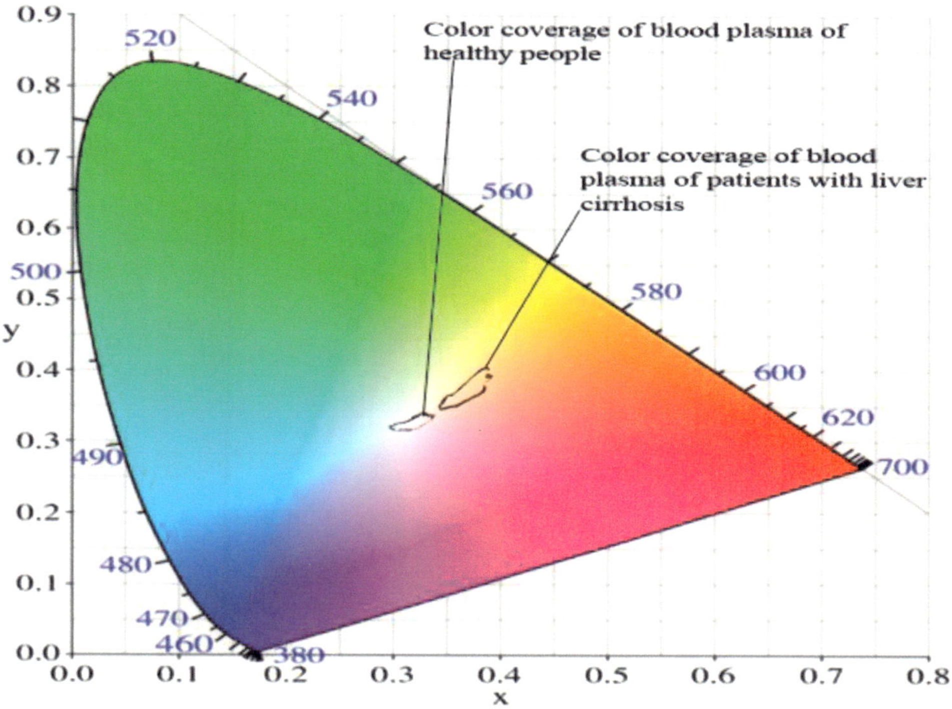

Figure 10.
Blood plasma of patients with liver cirrhosis and healthy people on the chromaticity diagram of system XYZ (Source C) [24, 25].

The color characteristics of the samples of blood plasma were calculated after processing spectra. The selective figures of the chromaticity coordinates of patients with liver cirrhosis and healthy people are shown in **Table 13**.

Then the statistical analysis of the data was made. The basic statistics for all the investigated samples are shown in **Table 14**.

Totalities of samples have a distribution close to normal and similar values of dispersion; therefore the t-test can be used to assess the reliability of the results. T of t-test for the chromaticity coordinate x was 10.57, for y—12.9. The critical value of t for confidence probability $p = 0.999$ is 3.5, which is much smaller than the obtained results. Consequently, the differences between chromaticity coordinates of groups of patients and donors were statistically significant.

The differences between the color characteristics of blood plasma of patients with liver cirrhosis and healthy subjects are shown in **Figure 10**.

Colorimetric method established that a healthy person is characterized by the following indicators of chromaticity coordinates:

$$x = 0.32 \pm 0.001, y = 0.32 \pm 0.002.$$

Patients with liver cirrhosis are characterized by the following color characteristics: $x = x = 0.352 \pm 0.006$ $y = 0.356 \pm 0.005$.

Having made the statistical processing, the data revealed that color characteristics of blood plasma of patients with liver cirrhosis differ from color characteristics of blood plasma of healthy people with a high degree of reliability.

Investigation of human blood plasma by colorimetric methods can be used to express diagnosis of liver cirrhosis. A healthy person is characterized by the following indicators of chromaticity coordinates: $x = 0.32 \pm 0.001$, $a = 0.32 \pm 0.002$.

Patients with cirrhosis of the liver are *characterized by the following color* characteristics: x = 0.352 ± 0.006 a = 0.356 ± 0.005.

Hence based on the integrated absorption spectra according to the standard CIE system, using the absorption coefficient for radiation in the visible wavelength range, we quantitatively determined the normal and pathological average color characteristics of human blood and its components (plasma and serum). The condition of the body is most adequately described using the lightness parameter for the aqueous solutions of plasma and serum. The method can be used in medical practice for rapid health assessment.

Author details

Mikhail Dolomatov
Department of Physical Electronics and Nanophysics, Physical and Technical Institute, Bashkir State University, Ufa, Bashkortostan Republic, Russia

*Address all correspondence to: mdolomatov@bk.ru

References

[1] Bensaude-Vincent B, Stengers I. A History of Chemistry. Cambridge, MA and London: Harvard University Press; 1997. 305 p. ISBN: 0-674-39659-6

[2] Dolomatov MY. Some physic-chemical aspects of prediction properties of multicomponent systems in conditions of extreme influences. Journal of D.I. Mendeleev Russian Chemical Society. 1990;**36**(5):632-639

[3] Dolomatov MY. Fragments of Real Substance Theory, From Hydrocarbon Systems to Galaxies. Moscow: Chemistry; 2005. p. 208. ISBN: 5-98109-018-9

[4] Dolomatov MY, Mukaeva GR, Shulyakovskaya DO. Electron phenomenological spectroscopy and its application in investigating complex substances in chemistry, nanotechnology and medicine. Journal of Materials Science and Engineering B. 2013;**3**(3):183-199

[5] Dolomatov MY, Kydyrgychova OT, Dolomatova LA, Kartasheva VV. Color the characteristics petrochemical systems. Journal of Applied Spectroscopy. 2000;**67**(3):536-540

[6] Dolomatov MY, Yarmukhametova GU. Correlation of color characteristics with conradson carbon residue and molecular weight of complex hydrocarbon media. Journal of Applied Spectroscopy. 2008;**75**(3):433-438

[7] Dolomatov MY, Jarmuhametova GU, Dolomatova LA. The interaction of color and physicochemical properties of hydrocarbon systems in colorimetric systems RGB and XYZ. Journal of Applied Physics. 2008;**4**:43-49. (Russia)

[8] Dolomatov MY, Shulyakovskaya DO, Yarmukhametova GU, Mukaeva GR. Evaluation of physico-chemical properties of hydrocarbon systems based on spectrum-property and color-property correlations. Chemistry and Technology of Fuels and Oils. 2013; **49**(3):273-280

[9] Dolomatov MY, Jarmuhametova GU. Number-average molecular weight definition of oils and petroleum residues by color characteristics. Chemistry and Technology Fuels and Oils. 2009;**4**: 46-49

[10] Dolomatov MY, Jarmuhametova GU. Calculation of average molecular mass, cocking capacity and viscous flow activation energy of oil surface samples by color-properties correlations. Geology, Geophysics and Reservoir Engineering. 2009;**7**:35-38. (Russia)

[11] ISO/CIE 10526. CIE standard illuminants for colorimetry

[12] Fairchild MD. Color Appearance Models. 2nd ed. Chichester, UK: Wiley-IS & T Series in Imaging Science and Technology; 2005

[13] The directory "Methods of research in textile chemistry". G.E. Krichevsski ed. Moscow: Lights industrial publishers; 1993

[14] Dolomatov MY, Mukaeva GR. Method for determining the ionization potential and electron affinity of atoms and molecules using electron spectroscopy. Journal of Applied Spectroscopy. 1992;**56**(4):344-347

[15] Dolomatov MY, Shulyakovskaya DO, Mukaeva GR, Jarmuhametova GU, Latypov KF. Simple characteristics estimation methods of material and molecule electronic structure. Journal of Materials Science and Engineering B. 2012;(4):261-268

[16] Dolomatov MY, Jarmuhametova GU, Shulyakovskaya DO. The estimation of the first ionization

potentials and electron affinity of the molecules of polycyclic organic semiconductors applying the color characteristics in colorimetric systems XYZ and RGB. Journal of Applied Physics. 2011;**1**:20-31. (Russia)

[17] Dolomatov MY, Shulyakovskaya DO, Mukaeva GR, Paymurzina NK. Testing amorphous, multi-component, organic dielectrics according to their electronic spectrums and color characteristics. Applied Physics Research. 2012;**3**:83-87

[18] Dolomatov MY, Mukaeva GR, Jarmuhametova GU, Shulyakovskaya DO. Simple definition methods of electron structures of materials and molecules for nanoelectronics. Nanotech Europe, Berlin, Germany; 2009. pp. 172. Available from: http://www.nanotech.net/content/conference/abstract-submission/simple-definition-methods-electron-structures-materials-and-m

[19] Dolomatov MY, Paymurzina NK, Latypov KF, Kovaleva EA. Specific quantum effects in atomic and molecular systems. Journal of Materials Science and Engineering A. 2013;**3**(12):183-190

[20] Dolomatov MY, Kovaleva EA, Paymurzina NH. Phenomenological approach to mathematical analysis of the electron spectra of organic semiconductors. Journal Nanotechnology: Development and Applications—XXI Century. 2016;**8**(4):43-48. (In Russian)

[21] Dolomatov MY, Latypov KF. Determination of heterocyclic molecules ionization potential based on optical absorption spectra of electromagnetic radiation in the visible and UV range. Fotonika (Photonika). 2017;**4**:60-67. (In Russian)

[22] Dolomatov MY, Kalashchenko NV, Dezortsev SV, Araslanov TR. Features of color characteristics of blood plasma of patients with liver cirrhosis in the colorimetric system XYZ as compared with healthy people. International Journal of Clinical Medicine. 2012;**3**(3):211-214

[23] Dolomatov MY, Kalashchenko NV, Arslanov TR, Dezortsev SV. Spectroscopic phenomenological estimation of the functional state of human organism in rate and pathology. International Journal of Clinical Medicine. 2011;**2**:79-81

[24] Kalashchenko NV, Dolomatov MY, Dezortsev SV. The Electronic Phenomenological Spectroscopy Blood of Human in Rate and Pathology, Theory and Practice Aspects. Moscow: Inter; 2010. p. 256. ISBN: 978-5-98761-034-3

[25] Kalashchenko NW, Dolomatov MY, Dezortsev SV, Popova EA, Kurmankaeva RR. Normal and pathological color characteristics of human blood components. Journal of Applied Spectroscopy. 2006;**3**:220-225

[26] Dolomatova LA, Kalashcenko NV, Dezorsev SN, Dolomatov MY. Definition of level of persons health on color features of the biological liquids in RGB colorimetric system. Journal Biomedical Radio Electronics. 2009;(6):10-13

[27] Wales DJ, editor. Intermolecular Forces and Clusters. Vol. 1. New York: Springer Berlin Heidelberg; 2005

[28] Jensen F. Introduction to Computational Chemistry. John Wiley& Sons; 2007. 599 p

[29] Tikhonov VI. Statistical Radiotechnics. Moscow: Radio i sviaz' Publ; 1982. 624 p. (In Russian)

[30] Dolomatov MY, Petrov AM, Bakhtizin RZ, et al. Asphaltenes as new objects for nanoelectronics. IOP

Conference Series: Materials Science and Engineering. 2017;**195**:29-32

[31] Kamyshnikov VS. Clinical Biochemical Laboratory Diagnostics: Handbook (in Russian). Interpresservis. Vols. 1 and 2. Minsk; 2003

[32] Barsegyants LO. Forensic Medical Investigation of Material Evidence (in Russian). Moscow: Meditsina; 1999

[33] Kudryashov LS, Gurinovich GV, Potipaeva NN. Method for meat quality control. Russian Federation Patent No. 2092836; 1997

[34] Kudryashov LS, Gurinovich GV. Molotchnaya Industriya, No. 5 (1998) (in Russian). Available from: http://food-machex.mtc.ru/mi/mi_05_98/topic7.htm

[35] Nazarenko GI, Kishkun AA. Clinical Assessment of Laboratory Research Results (in Russian). Moscow: Meditsina; 2000

[36] Pupkova VI. Determination of hemoglobin in blood (in Russian). Available from: http://www.vector-best.com/ru/brosh/hemo-glob.htm

[37] Berezov TT, editor. Laboratory Manual for Biological Chemistry (in Russian). Moscow: Meditsina; 1976

Chapter 3

Confocal Laser Scanning Microscopy for Spectroscopic Studies of Living Photosynthetic Cells

Natalia Grigoryeva and Ludmila Chistyakova

Abstract

Self-fluorescence of light-harvesting complex is a powerful tool for investigation of living photosynthetic microorganisms. As the physiological state of single cells of such microorganisms is closely related to the operation and activity of photosynthetic system, any variations in spectroscopic properties of their self-fluorescence indicate the changes in their physiological state. In this chapter, we present several applications of confocal laser scanning microscopy (CLSM) for investigation of living photosynthetic cells. A set of ordinary CLSM techniques will be applied for studying of cyanobacteria (or blue-green algae) such as 3D imaging, spectral imaging, microscopic spectroscopy, and fluorescence recovery after photobleaching (FRAP). Cyanobacteria were chosen as a model microorganism due to their great importance for different scientific and biotechnological applications. Cyanobacteria are the most ancient photosynthetic microorganisms on Earth. Nowadays, cyanobacteria are one of the most wide-spreaded organisms in nature, and the ecological aspect in their investigation is quite valuable. On the other hand, thousand strains belonging to different species are cultivated in biolaboratories all over the world for different biotechnological applications such as biofuel cells, food production, pharmaceuticals, fertilizers, etc. Thus, the noninvasive spectroscopic methods are quite important for monitoring of physiological state of cyanobacterial cultures and other photosynthetic microorganisms.

Keywords: confocal laser scanning microscopy, fluorescent microscopic spectroscopy, spectral imaging, 3D bio-imaging, cyanobacteria, blue-green algae, photosynthetic system, photoactive pigment, light-harvesting complex

1. Introduction

Confocal laser scanning microscopes (CLSMs) are distinguished by their high spatial and temporal resolution [1, 2]. Modern laser scanning microscopes are unique tools for visualizing cellular structures and analyzing dynamic processes inside single cells. They exceed classical light microscopes especially in their axial resolution, which enables to acquire optical sections (slices) of a specimen. An object can thus be imaged completely in three dimensions and subsequently visualized as a 3D computer image. Apart from simple imaging, confocal laser scanning microscopes are designed for the quantification and analysis of image-coded information. Among other things, they allow easy determination of fluorescence intensities, distances, areas, and their changes over time. In particular, they are capable of quickly detecting and quantitatively unmixing the spectral signatures of fluorescent objects. Many software functions analyze important parameters such as the degree of colocalization of labeled structures, or the ion concentration in a specimen. New acquisition CLSM tools include the detection of quantitative properties of the emitted light such as spectral signatures and fluorescence lifetimes. The most impressive feature of modern CLSMs is their capability for single-cell microscopic spectroscopy, which allows to obtain spectroscopic information inside single cells and small regions. Another group of applications is the quantitative investigations of dynamic processes in living cells using techniques such as fluorescence recovery after photobleaching (FRAP), fluorescence resonance energy transfer (FRET), photoactivation, and photoconversion.

The specific field of CLSM application is the investigation of self-fluorescence of living cells. The conventional biological fluorescent studies are based mostly on using fluorescent dyes and labels (chromophores). But it is well-known that the fluorescent dyes, even most flexible, affect living cells considerably. Thus self-fluorescence should become the most suitable tool for noninvasive investigation of changes in physiological state of living cells. For instance, the photosynthesis research employs the detection of self-fluorescence as a key method to study the metabolic mechanisms in photosynthetic cells and to detect photosynthetic efficiency in vivo. Recent rapid development of confocal microscopes functionality initiates new directions in subcellular biology research. However, the experiments with photosynthetic cells require some additional specific skills and techniques to perform measurements and to carry out data processing [3–7]. The efficiency of photosynthesis and photosynthetic rate are highly dependent on irradiance. This can be seen in the light-dependency of various photosynthetic parameters [8]. Moreover, not only light quantity, but also light “quality” (wavelength) is an important factor. Thus, special spectroscopic methods are required to study the physiology of phototrophic microorganisms [9]. These organisms employ light-dependent photosynthesis as the main energy source for their metabolism and the detected self-fluorescence finally reflects the diversity in morphological and physiological states of their photosynthetic cells [3–7].

CLSM single-cell microscopic spectroscopy is undoubtedly the most powerful tool for in vivo investigation of physiological processes in photosynthetic organisms (cyanobacteria, algae, and higher plants). The investigation of self-fluorescence of single living cells reveals the relation between the physiological state and the operational activity of photosynthetic system. A lot of interesting static and dynamic effects can be studied by means of confocal laser scanning microscopy. The investigation of self-fluorescence gives the information about single-cell processes as well as about the collaboration in cell communities. Changes in spectral characteristics of living photosynthetic cells indicate changes in their physiological state and can be applied for the studies of the results of stress states and external actions.

Recent progress in confocal laser scanning microscopy (CLSM) gives an opportunity to investigate different physiological processes in photosynthetic organisms on a single-cell level. Such CLSM applications as spectral unmixing and lambda-scanning provide the recording of spectral characteristics from living cells. FRAP and FLIP applications allow to study dynamic processes such as cell membrane fluidity and phycobilisome diffusion along thylakoid membrane. The nondestructive spectroscopic analysis conducting in vivo at a sub-cellular level allows to obtain more complete information about special features of individual cyanobacterial cells and supports the registration of very weak variations in their physiological state. For example, a novel technique for discrimination of cyanobacterial species and physiological states of the cells belonging to one strain was elaborated by the authors of the present chapter [3–5]. The technique is based on a strict relation between physiological state and genera affiliation of cyanobacterial cells and the intensity and the shape of corresponding single-cell fluorescence spectra, obtained by means of confocal microscopic spectroscopy. Light-, heat-, ultrasound- and toxin-induced changes can be distinguished by means of confocal microscopic spectroscopy since all these external actions are stress factors affecting photosynthetic process [5]. The application of such techniques for automation of on-line monitoring will give an additional opportunity to rise an effectiveness of biotechnological production and will bring in a valuable contribution to the development of innovative approaches in environmental monitoring [4, 5].

Here we present several experimental approaches to study the metabolic mechanisms in single photosynthetic cells in vivo. They are accompanied by several examples of in vivo investigations. Three main CLSM tools will be discussed in details: spectral imaging, fluorescent microscopic spectroscopy, and FRAP. All presented results were obtained using cyanobacterial strains from CALU collection of the Core Facility Center "Centre for Culture Collection of Microorganisms" of the Science Park of St. Petersburg State University as a model objects for CLSM studies.

2. Natural fluorescence

Opposite to the absorption spectra, the in vivo fluorescence spectra are much more informative. Fluorescence detection is a powerful tool owing to the existence of natural fluorescence from phycobilins and chlorophylls. It is a highly sensitive, nearly instantaneous, noninvasive way to study various components and processes in situ and in vivo. Although the fluorescence spectra contain the information only about photosynthetic apparatus of different algal groups, they include the information about the chemical structure of light harvesting complex (LHC) and accessory pigment-proteins, as well as about the character of links between pigment-protein complexes and the efficiency of energy transfer in the light harvesting process. When compared with absorption, fluorescence is affected by the excitation wavelength and energy. Thus, the use of different excitation wavelengths can provide more detailed information for the study of single-cell composition.

Today, there is no doubt that the in vivo analysis of fluorescence parameters of light-harvesting complexes is a powerful tool for studying the effect of a wide variety of environmental factors on photosynthetic organisms. The intensity of fluorescence emitted by single photosynthetic cells in vivo depends on the structure and operational effectiveness of photosynthetic apparatus, reflecting the individual characteristic of cyanobacterial strain and in-time physiological state of the cells under consideration. The environmental changes cause the changes in bioenergetic processes occurring in cyanobacterial cells, and so they significantly affect the kinetics parameters and spectral features of the intrinsic fluorescence of photosynthetic apparatus. Thus, the intrinsic fluorescence spectra of a particular type of cyanobacteria, the so-called "fluorescent fingerprints", can be used to identify photosynthetic pigments and to determine the viability of individual cells. These "fluorescent fingerprints" can be easily obtained by the routine lambda-scanning at most of confocal laser scanning microscopes.

2.1 Model object

Cyanobacteria, used in this study as a model object, are photoautotrophic prokaryotes. They are important microorganisms that contributed to the early oxygenation of the atmosphere and oceans on Earth 3.5 billion years ago. Now, cyanobacteria species are widely spreaded in nature and different morphologies with unicellular and filamentous forms can be found among them (**Figure 1**). Cyanobacteria as oxygen-evolving photosynthetic prokaryotes are good candidates for being a suitable system for numerous biological applications. Cyanobacteria have special metabolic features of autotrophic carbon and nitrogen assimilation and energy supply via photosynthesis, which presents an increased potential for the next generation of sustainable bioproduction.

In the last years, the investigation of taxonomy, physiology, morphology, and genetics of cyanobacteria attracts a considerable attention due to their potential application in biotechnology [10–16] and biosensing [17–19]. Cyanobacterial hydrogen has been considered as a very promising source of alternative energy and has now been made commercially available. Cyanobacteria are also used in aquaculture, wastewater treatment, food, fertilizers, agriculture, and production of secondary metabolites including exopolysaccharides, vitamins, toxins, enzymes, and pharmaceuticals. In addition, the ecological aspect of the harmful bloom monitoring and control makes an important contribution to this rising interest to cyanobacterial problem. A vast amount of different techniques were elaborated to achieve a nowadays insight of the physiological processes that rules cyanobacterial life and their genetic background.

As far as cyanobacteria are ancient photosynthetic microorganisms, they have complex and effective light harvesting apparatus, which exhibit multifarious self-fluorescent properties. These intrinsic natural features are essential for fluorescent non-destructive analysis on a single-cell level and give a unique opportunity for steady-state and time-resolved investigations of various physiological processes in vivo.

In these studies, several cyanobacterial strains from CALU collection of the Core Facility Center "Centre for Culture Collection of Microorganisms" of the Science Park of St. Petersburg State University were utilized.

Before the investigations were carried out, the cyanobacterial cells were cultivated during 7 days in BG 11 medium (at 28°C) under continuous white light irradiation (fluorescent tubes, 40 μmol photons/m^2/s). The cyanobacterial cells for visualization were taken at stationary phase of growth. For CLSM imaging, the

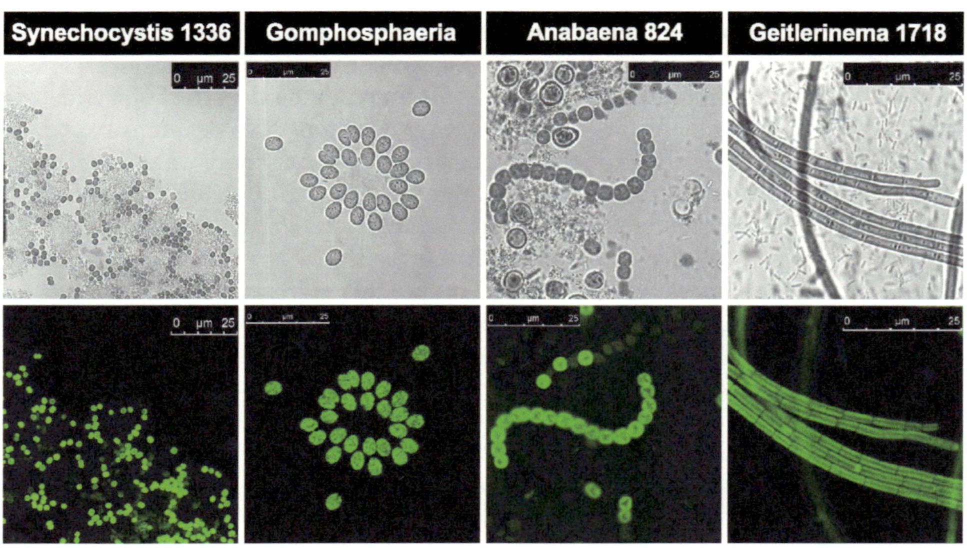

Figure 1.
Confocal laser scanning photomicrographs illustrating the morphological features and biological diversity of free-living and laboratory cyanobacterial strains: Synechocystis CALU 1336, Gomphosphaeria (wild type), Anabaena CALU 824, and Geitlerinema CALU 1718. Both fluorescence and transmission photomicrographs are presented. The white bar corresponds to 25 μm.

living cells were placed onto a glass slide and were allowed to settle. A glass coverslip was placed on top and sealed with nail polish. Samples were immediately imaged. All studies were performed with living cells at room temperature. Note here, that any centrifugation and/or resuspending of cyanobacterial cultures before microscopic measurements change considerably the physiological state of cells under consideration, and thus should be eliminated.

2.2 Light-harvesting system

Photosynthetic system of cyanobacteria, in contrast to higher plants, contains the external membrane light-harvesting complexes. Their antenna complex for photosystem II (PS II), and to some extent for photosystem I (PS I), is extrinsic and formed as a large multiprotein organelles, which are located on the stromal side of the thylakoid membranes. These supramolecular pigment-protein complexes are called phycobilisomes (PBSs). The detailed description of the morphology, structure, chemical, and optical properties of light-harvesting complex of cyanobacteria, phycobilisomes, and detached phycobilins can be found in numerous publications [8, 20–34]. Here, we only pointed out several main features that were essential for further discussion.

The main accessory pigments in cyanobacteria are phycobilins. The phycobilins which are bounded to proteins are known as phycobiliproteins. The three classes of phycobiliproteins in antenna complexes are allophycocyanin, phycocyanin, and phycoerythrin. However, in some cyanobacteria, phycoerythrin can be replaced by phycoerythrocyanin or both pigments can be lacking; phycocyanin and allophycocyanin are constitutively present in all cyanobacteria.

PBSs are assembled from 12 to 18 different types of polypeptides which may be grouped into three classes: (1) phycobiliproteins, (2) linker polypeptides, and (3) PBS-associated proteins. Phycobiliproteins, a colored family of water-soluble proteins bearing covalently attached, open-chain tetrapyrroles known as phycobilins. On the other hand, most of linker polypeptides do not bear chromophores.

Phycobilisomes are constructed from two main structural elements: a core substructure and peripheral rods that are arranged in a hemidiscoidal fashion around that core (**Figure 2**). The core of most hemidiscoidal phycobilisomes is composed of three (or two) cylindrical subassemblies, which are arranged side-by-side and form a triangle stack. Each core cylinder is made up of four disc-shaped phycobiliprotein trimmers, allophycocyanin (APC), allophycocyanin B (APC-B), and APC core-membrane linker complex (APC-LCM). By the core-membrane linkers, PBSs are attached on thylakoids and structurally coupled with PSII. The peripheral cylindrical rods (six or eight) radiate from the lateral surfaces of the core substructure and are usually not in contact with the thylakoid membrane. The rods are made up of hexamers, disc-shaped phycobiliproteins, phycoerythrin (PE), phycoerythrocyanin (PEC) and phycocyanin (PC), and corresponding rod linker polypeptides [8, 24–26, 32–35].

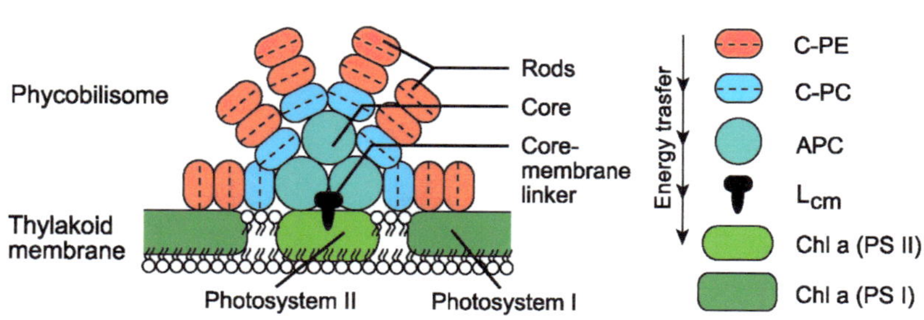

Figure 2.
Schematic drawing of a phycobilisome and the photosynthetic energy transfer to the reaction center.

2.3 Absorption, fluorescence, and energy transfer

As it was mentioned above, the fluorescence is one of the most powerful ways to probe photosynthetic systems, because it reports on the energy transfer and trapping. The intrinsic fluorescence of photosynthetic organisms originates from excited states that were trapped by light-harvesting system and lost before photochemistry took place.

In the light-harvesting pigment-protein complexes of cyanobacteria, the pigment molecules are excited by photons. In a nonradiating-induced resonance transfer process leading up to the reaction centers, they transfer the excitation energy to other pigment molecules which are excited in turn. In phycobilisomes, this fast process has an efficiency of almost 100%.

There are two mechanisms that serve the efficient excitation transfer in the light harvesting complex of cyanobacteria: the inductive resonance (Förster) transfer, applicable at long distances and weak interactions, and the occurrence of delocalized excitons, applicable at short distances and strong interactions [26].

The most likely mechanism by which excitation hops from one pigment complex to another across distances greater than several Angstroms is inductive resonance transfer, also known as Förster transfer [36, 37]. The interaction between an electronically excited pigment and its unexcited neighbors results in a downward transition of the initially excited group coupled with an upward transition in the nearby acceptor pigment by a through space Coulombic interaction (**Figure 3**). The energy donor and energy acceptor molecules must have an energy state in common, because when the excitation hops from donor to acceptor, the conservation of energy is required. This can be so only if the two molecules have a common energy state and therefore spectral transitions at the same wavelength. It should be noted, that the requirement for overlap of the fluorescence emission spectrum of the donor and the absorption spectrum of the acceptor does not lead to the process of emission of a photon by the donor, which is followed by absorption of a photon by the acceptor. The Förster transfer is a nonradiative process, which means that no photon emission or absorption is involved.

A critical step in the energy storage process is energy transfer between the antenna and the reaction center, where separation of the electron from a positively charged hole occurs. The molecules that accomplish the last event are organized in the photosynthetic membranes in a highly specific fashion to achieve the high efficiency of light energy conversion to photochemistry [30, 38].

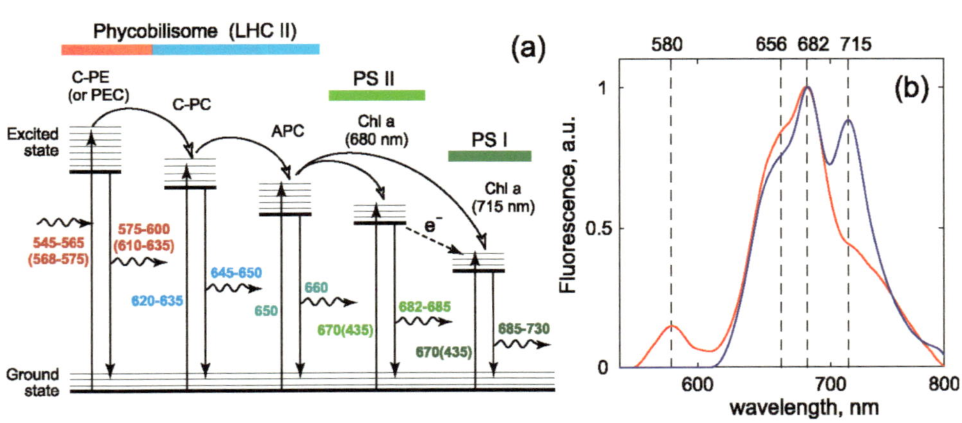

Figure 3.
Schematic illustration of the energy transfer in light-harvesting system of cyanobacteria (a). Panel (b) represents the normalized in vivo single-cell fluorescence emission spectra of two cyanobacterial species: blue line—Leptolyngbia CALU 1713 and red line—Nostoc CALU 1817. Excitation wavelength 488 nm. Dashed lines and numbers over them indicate emission wavelengths of PE, PC, and Chl a of PCII and PCI, correspondingly.

The more distal parts of the antenna system, often a peripheral antenna complex (phycobilisome), maximally absorb photons at shorter wavelengths (higher energies) than do the pigments in the antenna complexes that are proximal to the reaction center. Subsequent energy transfer processes are from these high-energy pigments physically distant from the reaction center to lower energy pigments that are physically closer to the reaction center (**Figure 3**). With each transfer, a small amount of energy is lost as heat, and the excitation is moved closer to the reaction center, where the energy is stored by photochemistry. Note, that the probability of excitation energy escapes from the trap in the form of fluorescence at all transfer steps is nonzero and depends on the intensity and wavelength of the excitation light.

During the energy transfer process, the occasional quenching of the absorbed light by fluorescence can occur and this becomes the essential property for fluorescent spectroscopy. It usually represents a small fraction of the excited states and diminishes in a functioning photosynthetic complex. Nevertheless, the fluorescence is an extremely informative quantity, because it reports on the energy transfer and trapping. Both steadystate and time-resolved fluorescence measurements are widely used methods for probing the organization and functional state of photosynthetic systems.

The unique spectroscopic properties of different cyanobacterial strains may become a promising fingerprints for practical and laboratory application [28]. The polypeptide composition of PBS varies widely among strains of cyanobacteria. However, it should be noted that the degree of PBS compositional variability, which reflects the ability of an organism to adapt to environmental changes, varies from strain to strain. Moreover, for a single strain, it also depends upon the environmental conditions such as nutrient availability, temperature, light quality, and light intensity.

The intact PBSs in cyanobacteria harvest sun light in the visible range from 400 to 750 nm and transfer the energy to the chlorophyll a (Chl a) of the photosystems PSII and PSI [24, 34, 35, 39]. When the single pigment-protein complexes aggregate in PBSs, their absorption bands are broadened due to the additional splitting of energy levels in the cause of interaction with other biliproteins and linker polypeptides. This leads to a more efficient harvesting of the light energy. Blue and red wavelengths of the visible light (around 440 and 675 nm) are mainly absorbed by cyclic tetrapyrroles, chlorophylls, combined in PSI and PSII, while green, yellow, and orange wavelengths (between 550 and 650 nm) are mostly absorbed by open-chained tetrapyrrole pigments, the phycobilins, composed in extramembranous antenna structures [24].

The fluorescence of intact living cyanobacterial cells is originated from the efficiency of the energy transfer between all components of the energy transfer chain including the final step, the delivery to PSII or PSI (**Figure 3(a)**). Each transfer step results in the spectrum shape as a peak or shoulder (**Figure 3(b)**). This is due to the fact that when phycobilisomes are bound to the thylakoid membrane, most of the energy from the last component of phycobilisome is channeled to chlorophylls in the thylakoid membrane and thus did not shade the fluorescence of the previous steps in energy transfer chain. In the course of the energy transfer from the initially photoexcited phycobiliprotein to the reaction center of photosystems PSI and PSII, fluorescence is emitted from almost every type of pigment and can be used as a probe to examine the mechanism of energy transfer within the light-harvesting system [8, 25, 26, 28].

A convenient way to monitor this energy transfer process is to irradiate a sample with light that is selectively absorbed by one set of pigments and then monitor fluorescence that originates from a different set of pigments. Obviously, if the energy transfer is taken place between pigments, the light absorbed by one set of pigments is emitted by another set differently depending on the excitation wavelength. This type of fluorescence excitation experiment can also be used to measure quantitatively the efficiency of energy transfer from one set of pigments to another [26]. Moreover, different species of cyanobacteria contain different accessory pigment-proteins and specific linker-proteins between them, therefore a set of fluorescence emission spectra excited by different wavelengths have its own unique shape for the cells of one strain and are quite distinguishable from other species and strains. Such sets of fluorescence emission spectra can be used for automatic differentiation of cyanobacterial species.

3. Confocal laser scanning microscopy (CLSM)

CLSM applications in biology and medicine predominantly employ fluorescence. Cellular structures can be specifically labeled with dyes (fluorescent dyes = fluorochromes or fluorophores) in various ways. But the cyanobacterial photosynthetic pigments, chlorophyll a and phycobilins, have an inherent fluorescence at 543 and 633 nm excitation and 590–800 nm emission and need no labelling for visualization. To target other specific elements, many fluorescent dyes and labels can be used. In this study, we use CdSe/ZnS quantum dots for imaging extracellular substances. It is also possible to use the transmission mode with conventional contrasting methods, such as differential interference contrast (DIC), as well as to overlay the transmission and confocal fluorescence images of the same specimen area.

In the present investigation, both Leica TCS-SP5 and Carl Zeiss LSM 750 were used for investigation of living cyanobacterial cells. For Leica TCS-SP5, laser power settings are as follows: 29% of Ar laser power was reflected onto sample with Acousto-Optical Tunable Filter (AOTF) and further power percentage for its laser lines was: 30% of 458 nm laser-line and 10% for all other lines. A 405-nm line of diode UV laser was reflected onto sample with 3%, and HeNe laser lines 543 and 633 nm were reflected with 10% and 2%, respectively. An acousto-optical beam splitter (AOBS) was used to transmit sample fluorescence to detector.

For 2D imaging, Leica TCS-SP5 was utilized. To rise the sensitivity and contrast of 2D images, they were recorded at 405 nm excitation wavelength (diode UV laser) and by Leica HyD hybrid detector, which strongly improves contrast in comparison to PMTs. HyD gain: 100%. The images of 1024 × 1024 and 2048 × 2048 pixels were collected with a 63× glycerol immersion lens (glycerol 80% H_2O) with a numeric aperture of 1.3 (objective HCX PL APO 63.0 × 1.30 GLYC 37°C UV) and with additional digital zoom factor 10–35. The fluorescence emission images were accompanied with the transmission images (in the parallel channel), collected by a transmission detector with the photomultiplier voltages ranged from 300 to 500 V. The images were recorded with a pinhole setting of 1 Airy unit (the inner light circle of the diffraction pattern of a point light source, corresponds to a diameter of 102.9 μm with the lens used (see [1])).

Carl Zeiss LSM 750 with LSM Software ZEN 2009 was used for 3D-imaging. The images were collected with EC Plan-Neofluar 100×/1.3 oil M27 and with additional digital zoom factor 2.86. XY-scan of 1024 × 1024 pixel images was performed with resolution 8-bit in two channels. One pixel corresponds to 29.7 × 29.7 μm. A multi-dimensional acquisition tool was used for recording z-stack of 80 sections with z-step 0.10 μm.

Fluorescence emission spectra of the intact cells were measured by Leica TCS-SP5 at 8 excitation wavelengths corresponding to all available laser lines. The excitation wavelengths are: 458, 476, 488, 496, 514 nm—the lines of Ar laser, 405 nm is the line of diode UV laser and 543, 633 nm are the lines of HeNe laser. Chlorophyll fluorescence was excited by the 405, 458, 476, and 488 nm laser lines, and phycobilisome (PBS) fluorescence was induced by the 496, 514, 543, and 633 nm laser lines. Fluorescence emission was detected for PE at 570–600 nm, for

PC and APC at 650–670 nm, and for Chl a of both photosystems at 675–720 nm. The whole emission spectrum between 520 and 785 nm was recorded using the lambda scan function of the "Leica Confocal Software" by sequentially acquiring a series ("stack") of 38–45 images, each with a 6-nm fluorescence detection bandwidth and with 6-nm wavelength step. For obtaining fluorescence-intensity information, images of 512 × 512 pixels were collected with objective HCX PL APO 63.0 × 1.30 GLYC 37°C UV and with additional digital zoom factor 5–9. One pixel corresponds to 53.5 × 53.5 nm. The photomultiplier (PMT) voltages were used in the range from 900 to 1100 V. For better signal yield, lambda scans were performed with "low speed" setting (400 Hz) in bidirectional scan mode and with a pinhole setting of 1 Airy unit. Regions of interest (ROIs) representing single cells or subcellular regions were used to calculate fluorescence spectra.

In CLSM applications, the laser light density in the focus point is high. Dwell time and the intervals between the illuminations may influence photo-damage and saturation of photosynthesis. Thus, since most chromophores and natural pigments bleach under the high laser excitation energies, a bleach-test should be performed [40]. During the detection, the fluorescence of the main accessory pigments for each cyanobacterial strain should be controlled and the changes in their fluorescence should not exceed 10–20%. The power of individual laser lines should be chosen according to the photodamage they cause. In our experiments, the repeated spectra were obtained under selected excitation power at a fixed point in a cell to check whether the excitation would affect the cells. In each case, it was shown that at the chosen excitation energies, the fluorescence spectra did not vary within the experimental error during 10–15 records. When excitation energy was increased, both the height and the center of the fluorescent peaks varied enormously with time because of photodamage or structure-breakdown in photosynthetic systems.

In the experiments, where several laser lines were involved for the investigation, the first spectrum was recorded again at the end of each series to control the initial state of the cell. To visualize differences between strains with higher spectral and spatial resolution, lambda scans were performed with 6-nm bandwidth and with 6-nm steps. As far as the fluorescence intensities depend on the excitation energy (which varies for different laser lines), sensitivity setting of the photomultiplier, and the distance from the sample, all spectra were usually normalized to their maximum, and only qualitative analysis was performed.

3.1 Spectral imaging and spectral unmixing

In a conventional light microscope, object-to image transformation takes place simultaneously and parallel for all object points. The specimen in CLSM is irradiated in a pointwise fashion, that is, serially, and the physical interaction between the laser light and the specimen fluorescence is measured point by point. To obtain information about the entire specimen, it is necessary to guide the laser beam across the specimen. Line-by-line scanning of the specimen is carried out with a focused laser beam deflected in the X and Y directions by means of two galvanometric scanners, and pixel-by-pixel detection of the fluorescence emitted by the scanned specimen details is performed by photomultiplier tube (PMT) or by more sensitive hybrid detector (HyD).

For examining flat specimens such as cell culture monolayers, it is usually sufficient to acquire one XY image to obtain the desired information. The same applies if the specimen is a three-dimensional tissue section of which a single optical section is representative. The thickness of the optical section (slice) and the focal position are selected so that the structures of interest are contained in the slice. The lateral resolution of a 2D image is defined by the pixel size in X and Y. The pixel size, in turn, varies with the objective used, the number of pixels per scan field, and the zoom factor. Pixels that are too large degrade resolution, whereas pixels too small require longer scanning times and thus bleach the specimen. The optimum pixel size for a given objective and a given zoom factor should be obtained for each special sample/experiment.

In this section, we examined changes in pigments of live, unfixed cells using spectral imaging. This technique captures an entire emission spectrum from every cell. The images corresponding to the four photosynthetic pigments have been pseudocolored for visualization purposes. The pseudocoloring is based on the colors shown in **Figure 5**: Chl a is red, PC-APC is green, and PE is light blue.

In **Figure 4**, images illustrated differences in fluorescence between diverse physiological states of Microcystis CALU 398 cells are presented. Two channels for fluorescence detection and one transmission channel were used. Channel 1 (a), colored false green, represents averaged fluorescence of PC and APC in the range 650–660 nm, channel 2 (b), colored false red, shows Chl a fluorescence at 678–710 nm, channel 3 (c) is a transmission microphoto, and in (d) all three channels are superimposed. Here small signal in both fluorescent channels corresponds to healthy cells, cells in bad physiological state (dying cells) have a high PC-fluorescence (green arrows), and dead cells have no fluorescence (white arrows). The high intensity in the shorter wavelength interval reflects that fluorescence in dying cells mainly originates from phycobilisomes at ca 650–660 nm. Healthy cells displayed weaker emission intensity over the whole spectral range, located at the periphery of the cells. This response correlates with effective light utilization and energy balance [26].

Figure 5 represents confocal laser scanning photomicrographs illustrating photosynthetic pigment localization in filamentous cyanobacterium Phormidium CALU 624. False color overlay images were obtained by simultaneous three channel

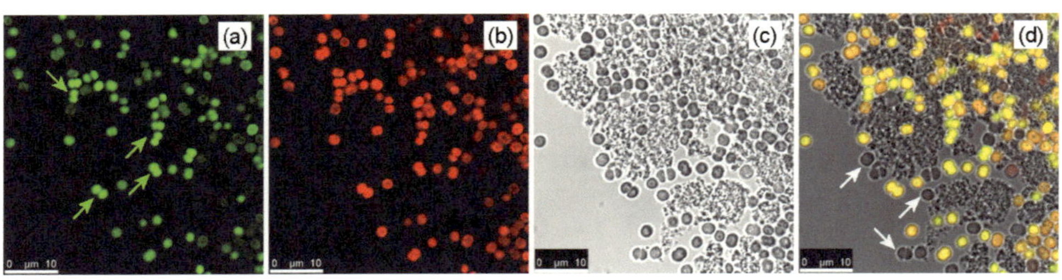

Figure 4.
CLSM micrographs of cyanobacterium Synechocystis 1336, visualized with CLSM. Green channel (a) represents fluorescence from phycobilins at 656 nm, red channel (b) from chlorophyll a at 682 nm, (c) represents transmission image and an overlap of all three is shown in (d). Scale bar = 10 μm.

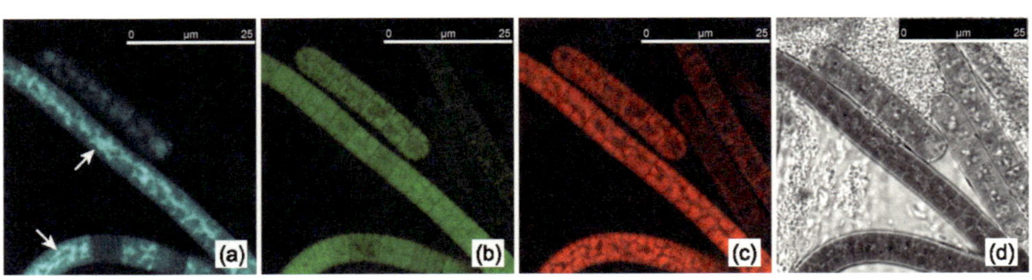

Figure 5.
False color overlay images of Phormidium CALU 624 cells at stationary phase of growth are visualized by CLSM simultaneously in three fluorescence detection channels. (a) Channel 1 (light blue): fluorescence signal from PE (575–585 nm), (b) channel 2 (green): PC-APC (650–660 nm), (c) channel 3 (red): Chl a (678–710 nm), and (d) channel 4: transmission microphoto. Scale bar = 25 μm.

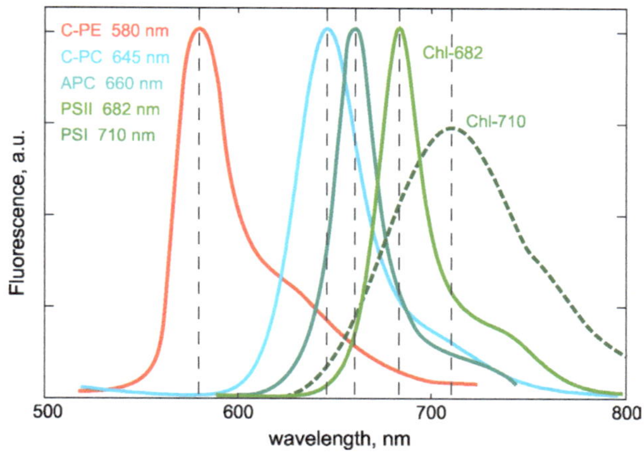

Figure 6.
Fluorescence spectra of individual phycobiliproteins, which can be used as reference data for spectral unmixing. Spectra were normalized to their maximum, and all values were adapted to our measurement conditions. The dashed lines indicate the respective fluorescence maxima (PE—580 nm; PC—645 nm; APC—660 nm; Chl (PSII)—682 nm; Chl (PSII)—710 nm).

fluorescence detection. Fluorescence images (**Figure 5(a)–(c)**) were obtained at different emission wavelengths, excited by a 405 nm laser beam. The fluorescence signal was collected at 647–652, 658–663, and 675–680 nm, and the false colors light blue, green, and red in **Figure 5(a)–(c)** represent the fluorescence emitted by PE, PC-APC, and Chl a, respectively. The highest intensity signal of PE-fluorescence is located in the center of the cells (white arrows), and the strongest signal from Chl a fluorescence is located at the cell periphery, the latter provides the evidence of the thylakoid arrangement.

Fluorescence spectra of living cyanobacterial cells are composed of overlapping fluorescence emission spectra from several photosynthetic pigments. The application of spectral unmixing with reference spectra for individual pigments (**Figure 6**) can give more pronounced and detailed images of pigment localization. Spectral unmixing can also be used to calculate the relative fluorescence shares of the individual pigments contributing to the spectrum.

Spectral unmixing is a method for the complete separation (unmixing) images with overlapping emission spectra. It is used with specimens labeled with more than one fluorescent dye, exhibiting excitation and emission crosstalk or with self-fluorescent specimens contained of several photosynthetic pigments (as in our case).

If we regard a pixel of a lambda stack that represents a locus in the specimen where several fluorescent pigments with their known reference spectra overlap, the cumulative measured spectrum can be expressed as a linear combination of the reference spectra multiplied by the corresponding intensity for each pigment. By means of known reference spectra, this equation can be solved for the intensities of each pigment. The reference spectra can either be loaded from a spectra database, or from literature [40–44], or directly extracted from the lambda stack. Obtained linear unmixing function will generate a multi-channel image, in which each channel represents only one pigment. The accuracy of the technique allows the complete unmixing even of such dyes or self-fluorescence pigments whose spectra have almost identical emission maxima.

Unfortunately, in living cyanobacterial cells, it meets some difficulties. The problems are caused by the fact that the light absorption and emission properties of isolated phycobiliproteins are rather different from those of the intact phycobilisomes in the living cyanobacterial cells. In living cells, the spectral properties

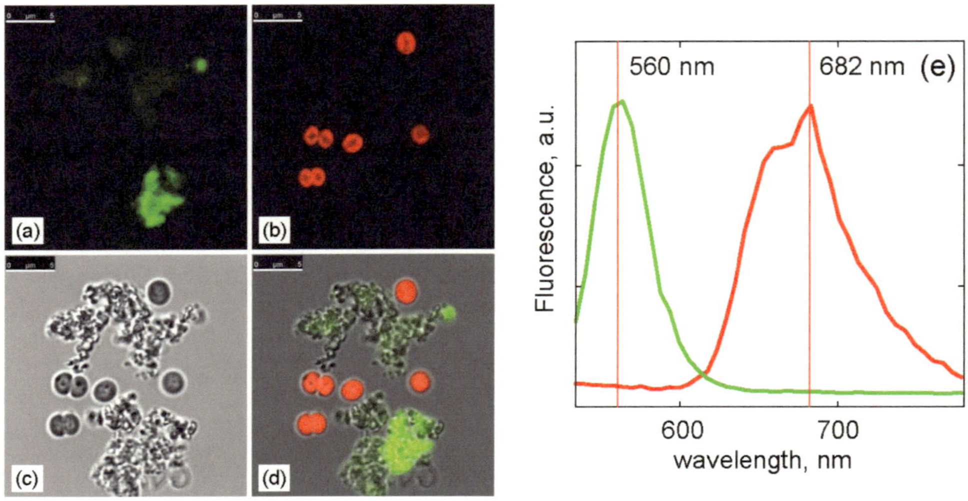

Figure 7.
Simultaneous observation of self-fluorescent photosynthetic cells of Microcystis CALU 398 and labeled by quantum dots extracellular polysaccharide. (a) Channel 1: quantum dots fluorescence (green), (b) channel 2: natural pigment fluorescence (red), (d) channel 3: corresponding transmission image, (d) all three channels superimposed. Fluorescence spectra in (e) have the colors corresponding to (a) and (b). Scale bar = 5 μm.

of pigments from certain organisms may differ crucially from the properties of the dissolved ones, for example, spectra of the components can vary in peak widths and may be shifted in wavelength due to the different pigment-protein and linker connections. In this case, only hyperspectral CLSM measurements [42] can give suitable reference spectra for decomposition.

If the overlapping of the emission spectra of the fluorescent components is not very much, then the image spectral unmixing can be easily obtained without any calculations. The example of such unmixing is presented in **Figure 7**. In **Figure 7**, the usage of combination of natural pigment fluorescence and fluorescence probes for imaging of cyanobacteria and their extracellular polymeric substances (polysaccharides) is demonstrated. To target extracellular substances, fluorescent labels should be used. Here, nonfluorescent polysaccharides were labeled by quantum dots CdSe/ZnS. Quantum dots have a fluorescence in green (520–580 nm) when excited by ultraviolet light (405 nm) and photosynthetic pigments of Microcystis CALU 398 have a fluorescence in red (620–780 nm). Thus the fluorescent signals will not overlap. The reference spectra for quantum dots and Microcystis CALU 398 cells were obtained by means of lambda scanning (**Figure 7(e)**).

3.2 3D bio-imaging

In addition to the possibility to observe a single plane (or slice) of a thick specimen in good contrast, optical sectioning allows a great number of slices to be cut and recorded at different planes of the specimen, with the specimen being moved along the optical axis (Z) by controlled increments. The result is a 3D data set, which provides information about the spatial structure of the object. The quality and accuracy of this information depend on the thickness of the slice and on the spacing between successive slices.

Once a 3D stack of images has been recorded, the user has various presentation options. The data may be displayed as a gallery of depth-coded images or as orthogonal projections of the XY, XZ, and YZ planes. To create a 3D impression on a 2D monitor, animations of different viewing angles versus time, 3D reconstruction, shadow projections, and surface rendering techniques are possible.

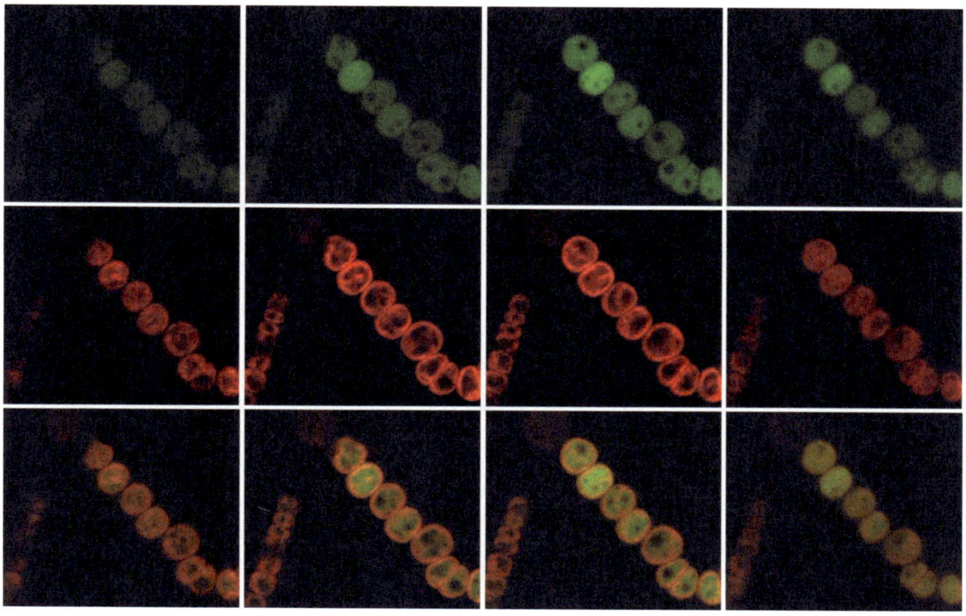

Figure 8.
Pigment localization through the depth. Two-dimensional sections of Anabaena CALU 824. Green channel: PE-fluorescence, red channel: Chl a-fluorescence, and bottom panel: two channels superimposed.

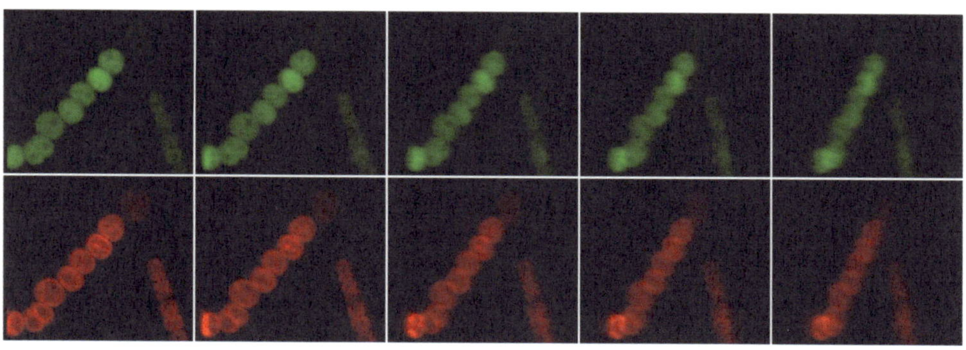

Figure 9.
Several frames from animation of 3D-model at different viewing angles. Two colors correspond to two fluorescent channels: green channel—PE-fluorescence and red channel—Chl a-fluorescence.

In **Figure 8**, several two-dimensional optical sections of a filament of Anabaena CALU 824, obtained by Carl Zeiss LSM 710 with LSM Software ZEN 2009, are shown. A series of XY images (z-stack), acquired in different focus positions represents the Z dimension of the specimen. Here we present images from two channels: green—PE-fluorescence, red—Chl a-fluorescence, and the bottom panel—two channels superimposed. To calculate 3D model from z-stack, much more sections should be acquired.

Figure 9 shows several frames from 3D reconstruction animation computed from a 3D data set in ImageJ 1.46r software (http://imagej.net/). To render 3D-model z-stack containing 80 images were recorded with 0.1 μm increment along z-axis.

3.3 Single-cell spectroscopy (lambda scanning)

Fluorescence spectra of living photosynthetic cells can be reliably analyzed by a microspectrofluorometric method (CLSM spectroscopy), which is implemented in many modern CLSMs. This method allows the precise localization of the fluorescence signal, even within a single living cell. Modern fluorescence microscopic spectroscopy or confocal laser scanning microscopic spectroscopy provides a unique opportunity to obtain the intrinsic fluorescence emission spectra from cyanobacterial cells in vivo [40–43, 45–46]. Moreover, using spectral unmixing, the fluorescence of individual spectral components can be resolved, and their relative intensities can be calculated [1–2].

The acquisition of spectral data becomes necessary when the cellular parameter to be measured is coded by changes of the emission spectrum. PMT element registers sequentially a different part of the spectrum, each part having a spectral width of 6 nm with a step 6 nm. The result is a lambda stack of XY images in which each image represents a different spectral window. Lambda stack is a series of images showing the same area, but in different spectral windows. From a lambda stack, the intensity of the signal for each pixel of the image can be extracted as a function of wavelength. These spectral "fingerprints" can easily be obtained for any image area by means of the mean of ROI function.

Although, the pigment structure of different cyanobacterial strains has been intensively investigated, the variations in self-fluorescent spectra of single cells for different cyanobacterial species has not been analyzed yet. We suppose that the best way to investigate the in vivo operation of photosynthetic system is a single-cell fluorescence microscopic spectroscopy. Single-cell detection can provide the information on small peculiarities that is regularly buried in normal ensemble average experiments. This is thus a good way to study the time evolution process and spectroscopic properties of individual cells. Both steady-state and time-resolved fluorescence measurements can be used for probing the intrinsic self-fluorescence by means of CLSM.

As it was mentioned above, phycobilisomes contains several kinds of biliproteins, and their fluorescence spectra reflect the contribution of each (**Figure 3**). Moreover, depending on the excitation wavelength, the room temperature fluorescence emission spectrum of intact cyanobacterial cells exhibits various extents of contribution of phycobilisome emission to the spectrum. If one exclusively excites Chl a, using a 458-nm line of an Ar laser, the emission spectrum by cyanobacterial cells shows no appreciable emission of PC or APC. This is indicated by PSI emission band at 715 nm and PSII emission band at 682 nm (**Figure 10**). The excitation by intermediate (blue and green) wavelengths (405, 488, and 496 nm) reveals fluorescent maxima of all photosynthetic pigments, as the light in this range is absorbed by all pigment-protein complexes almost in equal portions, and fluorescence emits by all steps of energy transfer chain (**Figure 3**). The direct excitation of cells in the PC absorption region at 514 and 543 nm results in emission spectrum with two main peaks at 580 and 656 nm, which are due to PE, PC, and APC emission. The spectra of the 633 nm excitation directly gives a prominent emission band at 656 nm, that originates from PC, omitting band at 580 nm, which cannot be excited by 633 nm, even for species that have PE. Other small emission bands, corresponding to fine pigment structure of antenna complex, are not resolved at room temperature. Comparative analysis of the series of fluorescence spectra for different cyanobacterial species and strains reveals visible variations in their shape (**Figure 10**). If the fluorescence spectra were taken from alive cells in normal physiological state, which are cultured in the same growth environmental conditions, then the interspecies variations in pigment/Chl a ratios are more pronounced than variations within the individual species. And species/strains differentiation could be carried out on the base of fluorescence analysis [3, 6–7].

Figure 10 shows the experimental sets of single-cell fluorescence spectra for Microcystis CALU 398, Merismopedia CALU 666, Leptolyngbya CALU 1715,

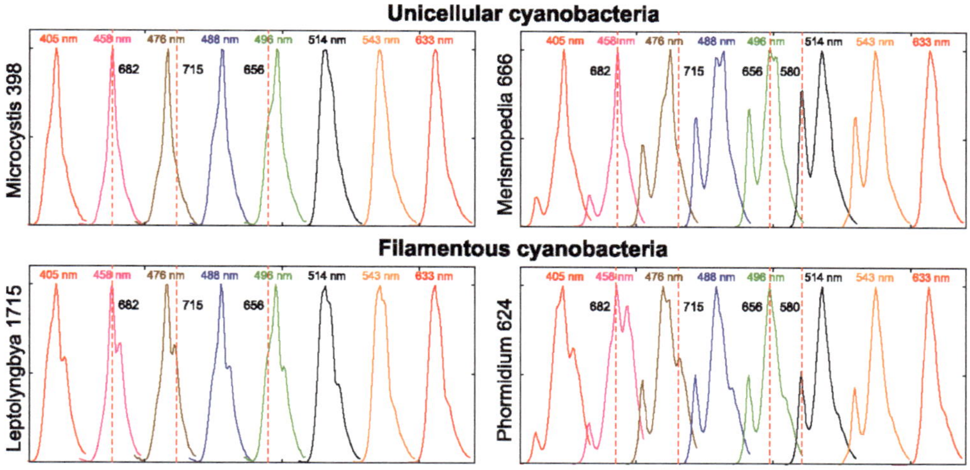

Figure 10.
Single-cell fluorescence spectra of different cyanobacterial strains (unicellular and filamentous): Microcystis 398, Merismopedia 666, Phormidium 624, and Leptolyngbya 1715. The excitation wavelengths (405, 458, 476, 488, 496, 514, 543, and 633 nm) are given over the curves. All spectra are normalized to the maximum intensity and shifted along X-axis for convenience of observation. The dashed lines indicate the fluorescence maxima of the individual pigments (PE—580 nm; PC—656 nm; Chl a—682, 715 nm).

and Phormidium CALU 624 (cyanobacterial strains are labeled according to CALU collection of the Core Facility Center "Centre for Culture Collection of Microorganisms" of the Science Park of St. Petersburg State University). Each spectrum set was obtained by means of confocal laser scanning microscope (CLSM) Leica TCS-SP5, using different laser-lines for excitation (405, 458, 476, 488, 496, 514, 543, and 633 nm). Corresponding excitation wavelengths are given over each spectrum. All spectra are normalized to the maximum intensity and shifted along X-axis for convenience of observation.

Fluorescence of photosynthetic pigments in the intact cells is affected by physicochemical and physiological processes that occur within and across the thylakoid membranes. The structure and function of the thylakoid membrane can alter under stress conditions. The alterations may be both short- and long-term, depending on the nature and duration of the stress. The dynamical changes in fluorescent spectra of living cyanobacterial cells in different stress states can be detected via time-resolved CLSM measurements and can be used for estimation of the physiological state and viability of cyanobacterial cultures at different experimental stages.

Figure 11 illustrates the temporal changes of in vivo fluorescence spectrum taking place in the living cell of cyanobacteria strain Synechocystis CALU 1336 under high-light and heat stress. The excitation wavelength in this experiment was 488 nm. The set of spectra was obtained from a single cell during 45 min with 1 min step under over-excitation conditions (only every fourth spectra is presented in **Figure 11(a)**). At first, several minutes (1–12 min) spectrum changes slightly and mostly in the range of PC-fluorescence (around 656 nm). Since 488 nm light is absorbed primarily by PC and Chl a, photosynthesis and energy transfer chain are quickly saturated and reaction centres closes, thus the fluorescence around 656 nm starts growing. During the next few minutes (12-28 min) a rapid growth of PC-APC fluorescence indicates that over-excitation leads to the uncoupling of phycobilisomes from thylakoid membrane, while the Chl a fluorescence remains almost the same during this time period (see **Figure 11(b)**). Small rise of the Chl a fluorescence intensity at 20–39 min can be explained by strong overlapping of Chl a and PC-APC emission spectra, which leads to a crosstalk between these fluorescent signals. Then, after 30 min of irradiation, the degradation of detached phycobilisomes and PSII

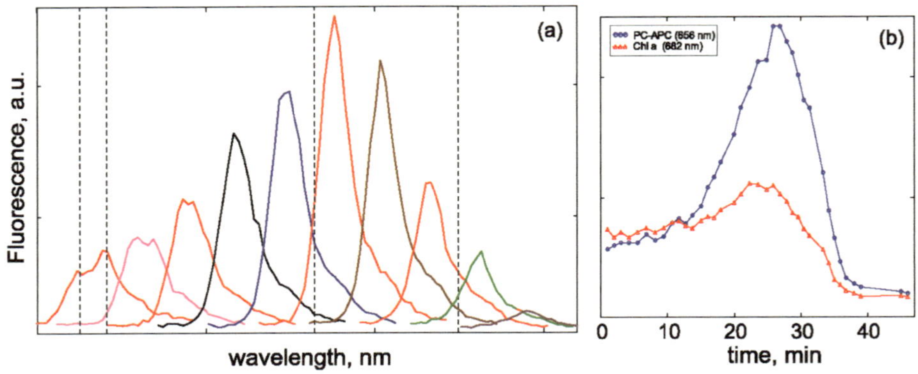

Figure 11.
Time degradation of living cell of cyanobacterial strain Synechocystis CALU 1336 under high-light and heat stress. (a) Spectra recorded at the excitation wavelength 488 nm with the time-step 4 min. All spectra are shifted along x-axis for convenience of observation. (b) Time dependence of the fluorescence intensity at 656 nm (PC-APC fluorescence) and at 682 nm (Chl a fluorescence).

pigment-protein complexes begins. The intensity of Chl a and PC-APC fluorescence decreases, and maximum is shifted to shorter wavelengths.

Two presented examples of CLSM lambda-scan application show the unique abilities of CLSM microscopic spectroscopy for investigation of physiological processes in cyanobacterial cells. More examples can be found in [3–5, 47].

3.4 Using fluorescence recovery after photobleaching (FRAP)

A field of growing importance is the investigation of living specimens that show dynamic changes. Dynamic processes in living specimens can be recorded by means of time series implemented in modern CLSMs. Data once acquired can be analyzed "off-line," that is, after image acquisition. Time series are defined by a start time and the time interval between two successive images. To analyze a time series, some option allows fluorescence intensity changes to be quantified in defined regions of interest (ROIs).

Within a time series, the CLSM permits selective, point-accurate illumination of ROIs with laser light. This function is useful for generating a photobleaching routine, for example, within a FRAP experiment (fluorescence recovery after photobleaching), for analyzing dynamic processes. Complex time series experiments, with different images to be taken at different sites within a specimen according to a defined time pattern, can be carried out by means of CLSM software.

Fluorescence recovery after photobleaching (FRAP) is a technique widely used in cell biology to observe the dynamics of biological systems. In photosynthetic organisms, it can be directly used, for example, for investigation of the dynamics of thylakoid membranes, including the diffusion of membrane components [39, 48–52].

Diffusion of membrane components is involved in a number of processes, for example, in investigation if the chromatic adaptation of light-harvesting apparatus, in describing of the electron transport in photosynthetic membranes and in studies of the biogenesis, turnover, and repair of the photosystems. Diffusion coefficients for certain thylakoid membrane components can also be estimated by indirect methods [53, 54], but FRAP and related optical techniques may be used for direct observation of the diffusion in biological systems.

The component whose diffusion is to be observed must be bleached by high laser power with the confocal spot scanning briefly over a small area of the sample. After bleaching, the laser power is decreased, and the whole sample is imaged. The bleached area of the sample will be seen as a dark, nonfluorescent patch on the image. Then the sample is repeatedly imaged, and if the bleaching will change somehow during the time, the diffusion coefficient can be derived from these characteristic changes [48, 49]. Usually the component whose diffusion is to be observed must be tagged with a fluorophore, but, as it was mentioned above, the great advantage of photosynthetic membranes is that their protein complexes are naturally fluorescent, so their diffusion can be observed without any necessity for specific fluorophore binding or GFP gene fusions. The only requirements for quantitative FRAP are that the membrane geometry is known, and the membrane environment is uniform over an area considerably larger than the area of the bleach. Unfortunately, most green plant thylakoid membranes have an intricate, convoluted structure, and extensive lateral heterogeneity on small scales. On the other hand, some cyanobacterial species have the membrane systems located in the periphery of the cell and are uniform enough in small regions. Anyway, by means of FRAP technique, it is always possible to see if particular components are mobile or not, to make a rough estimation of time-scales and to investigate a number of factors that affect their diffusion coefficient.

In contrast to typical green plants, in most cyanobacterial strains, there is no thylakoid membrane stacking and no extensive lateral heterogeneity. The preferred model organism used in previously published papers [39, 48, 49] is the cyanobacterium Synechococcus sp. PCC7942, which has suitable membrane conformation and is also well-characterized and transformable. Its cells can be elongated up to 10–20 μm by means of some growth techniques to achieve lateral heterogeneity in relatively large region. But in this work, we demonstrate that reasonable results can be obtained even for spherical cells of Microcystis firma CALU 398.

It is well known that phycobilisomes diffuse rapidly on the surface of the thylakoid membrane, while PS II reaction centers are normally almost immobile. Here we used FRAP to measure the mobility of phycobilisomes in the intact cyanobacterial cells. These experiments indicate that phycobilisomes may frequently decouple from reaction centers and diffuse randomly before attaching to another reaction center.

FRAP measurements were carried out with a laser-scanning confocal microscope Leica TCS-SP5. During the measurement, a series of images of the cell is recorded, and it is important to keep the laser power low enough not to cause significant further bleaching of the cell during repeated imaging. The basic procedure for the measurement is to record an image of the sample cell before bleaching (**Figure 12(a)**(1). Then the laser power is increased by a factor of 8–10, and the confocal spot is scanned only over a restricted area of the sample (light-blue vertical rectangle) to bleach out most of the fluorescence in that area. The laser power is then reduced again, and a series of post-bleach images is recorded. The best time-scale for the measurement depends on the rate of diffusion and has to be found by trial and error. In thylakoid membranes, the diffusion of both lipids and proteins is generally rather slow and a 1 s bleach, followed by a series of images recorded at 1.3 s intervals, was adequate to capture diffusion.

Figure 12 summarizes the steps involved in recording and analyzing FRAP data. In the presented experiment, a one-dimensional diffusion of phycobilisomes was measured in a cell of the cyanobacterium Microcystis firma CALU 398. Confocal laser scanning microscope Leica TCS-SP5 with immersion objective HCX PL APO 63.0x1.30 GLYC 37°C UV was utilized in this investigation. The He-Ne laser (633 nm, 150 mW) was used for excitation in both bleaching and scanning modes, and the laser power was 25 and 2% correspondingly. The vertical (Z) resolution was 0.2 μm, and the resolution in the x-y plane was about 24.1 nm. Digital zoom was 20. The laser was scanned over the sample with oscillating mirrors. Scan speed was 400 Hz. Fluorescence from the sample was separated from the excitation light with a beam-splitter, passed through a 102.9 μm pinhole and detected with a photomultiplier with PMT voltage 1000 V. The emission band for recorded images was 673–678 nm.

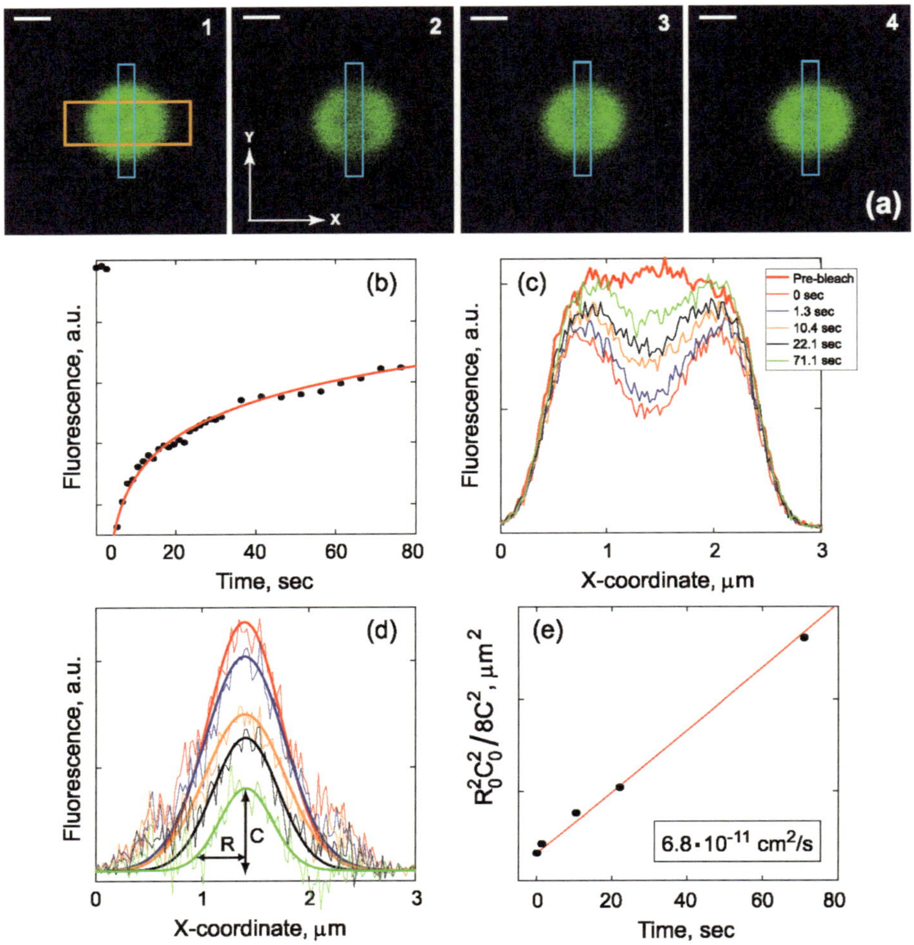

Figure 12.
Extracting data from a FRAP experiment on a cell of Microcystis CALU 398. (a) Fluorescence intensity images from FRAP-sequence, taken at −5, 0, 10.4, and 71.1 s, correspondingly. ROI 1 (light blue rectangular): bleaching area and ROI 2 (orange rectangular): the area for controlling fluorescence intensity profile across bleaching region. Scale bar = 1 µm. (b) Total fluorescence intensity of bleached area as a function of time. The recovery of the fluorescence is presented as open circles and exponential fitting function—solid line. (c) Pre-bleach and post-bleach fluorescence intensity profiles across bleaching region at t = 0, 1.3, 10.4, 22.1, and 71.1 s. (d) The subtracted post-bleach and pre-bleach fluorescence intensity profiles and corresponding Gaussian fitting curves at five time-points indicated at (c). (d) A plot of C-function for one-dimensional diffusion versus time. The obtained diffusion coefficient is indicated on the plot.

The first image (**Figure 12(a)**(1) was recorded prior to bleaching, by scanning the confocal spot in the XY mode. Excitation wavelength was at 633 nm, and the detection range was from 650 to 750 nm. The scan was then switched to another mode, the laser power was increased by a factor of 12, and the sample was bleached by scanning the laser repeatedly in one dimension across the cell for 1 s. Light-blue vertical rectangle indicates bleaching area. Then the laser power was reduced again to the initial value, the scan was switched back to the XY mode, and the cell was imaged by scanning over a square of 12.30 × 12.30 µm in the x-y plane (512 × 512 pixels) (**Figure 12(a)**(2–3). There was no detectable photobleaching during the recording of successive image scans. A series of 20 further images were recorded at time intervals 1.3 s and then another 10 images at time intervals 5 s. Images (a3) and (a4) show the state at 20th and 30th point (10.2 and 71.1 s, correspondingly). Within a few minutes, the bleaching profile spreads, becoming broader and shallower. This shows that phycobilisomes are diffusing, with unbleached phycobilisomes diffusing into the bleached area.

Leica software (LAS AF Lite 2.6) was used to extract the fluorescence profiles across the bleaching area (orange rectangle) from the images, summing all the pixel values in the Y-direction for each X coordinate. And then MathCad 14 was used for curve fitting and further calculations.

(**Figure 12(b)**) shows a typical time dependence of fluorescence recovery after photobleaching. The open circles denote the integrated signals over the bleached area, marked by the light-blue vertical rectangle in the fluorescence intensity images (a), and the line represents a fitting curve. First three points correspond to pre-bleaching images. Then goes a 1-s gap corresponds to bleaching period. The next 20 points were recorded with step 1.3 s and another 10 points with step 5 s.

One pre-bleach and five post-bleach fluorescence profiles across the bleaching area (orange rectangle) at a series of time-points were aligned (**Figure 12(c)**). The post-bleach profiles (shown in thin lines) were corrected by subtracting the pre-bleach profile (shown in thick red line), and the Gaussian curves were then fitted to the corrected fluorescence profiles (**Figure 12(d)**). The fitted Gaussian curves were used to obtain corresponding depth (C) and the half-width (1/e2) (R) of the bleach. Note that the bleaching profile remains Gaussian, but becomes broader and shallower with time. The change in C and R with time and diffusion distance in one dimensional diffusion model can be described according to the equation given in [49]. Finally, a suitable function of C is plotted versus time in (**Figure 12(e)**) and the corresponding diffusion coefficient was obtained. Here, R0 and C0 are the values of R and C at time just after bleaching. For one-dimensional diffusion, the plot shown in (**Figure 12(e)**) should be linear with gradient equal to the diffusion coefficient. The calculations for the diffusion coefficient of phycobilisomes give a result of 6.8 × 10–11 cm^2/s, which is similar to those given in [49]. Together with the linearity of the plot in **Figure 12(e)**, this indicates that this procedure can be applied to spherical cells also and that our measurement does not greatly perturb the system.

4. Conclusion

Light microscopy has been used for studying cells for many years and has advanced our understanding of key cellular processes. However, fixation involves nonphysiological procedures and only provides a snapshot view of cells at a single point in time. To truly understand cellular function, we need to extend our imaging capabilities in ways that enable us to follow sequential events in real time, monitor the kinetics of dynamic processes, and record sensitive or transient events. And CLSM gives such opportunity. In addition, CLSM technique allows the investigation not only cultivable, but also the noncultivable phototrophic microorganisms.

This study presents simple and practical advices for performing special confocal microscopy applications. Our approach allows to study detailed mechanisms of photoprotection and stress reactions in different cyanobacterial species. Fluorescence spectra of cyanobacterial photosynthetic pigments are easily recorded by spectral CLSM. The fluorescence shares of individual phycobiliproteins can be reliably determined by spectral unmixing, showing that the spectral resolution of CLSM is well suited for this approach.

Finally, with the advent of live cell imaging and the development of high- and superresolution technologies, it is now possible to acquire data on viable cells in a biologically relevant context providing us with a greater insight of cellular function than has previously been possible.

Note here, that there are a lot of additional techniques already implemented in modern CLSMs, which open new perspectives for single-cell investigation, such as: white laser, which provides the ability to obtain not only fluorescence emission spectra, but also single-cell excitation and absorption spectra [55]; hyperspectral CLSM which allows more precise fluorescence spectra through the cell thickness and gives more detailed fluorescent pigments location [42]; STED and multi-photonic techniques, that extend the CLSM abilities to single-molecular studies [56, 57]. They are the subject for further investigations.

Author details

Natalia Grigoryeva[1*] and Ludmila Chistyakova[2]

1 Saint-Petersburg Research Center for Ecological Safety, Russian Academy of Sciences, Saint-Petersburg, Russia

2 Saint-Petersburg State University, Saint-Petersburg, Russia

*Address all correspondence to: renes3@mail.ru

References

[1] Pawley JB, editor. Handbook of Biological Confocal Microscopy. Boston, MA: Springer; 1995. 346 p. DOI: 10.1007/978-1-4757-5348-6

[2] Wilson T. Confocal Microscopy. London: Academic Press; 1990. 426 p. DOI: 10.1007/978-1-4615-7133-9

[3] Grigoryeva N, Chistyakova L. Fluorescence microscopic spectroscopy for investigation and monitoring of biological diversity and physiological state of cyanobacterial cultures. In: Tiwari A, editor. Cyanobacteria. London: IntechOpen; 2018. pp. 11-44. DOI: 10.5772/intechopen.78044

[4] Rumyantsev VA, Grigor'eva NY, Chistyakova LV. Study of changes in the physiological state of cyanobacteria caused by weak ultrasonic treatment. Doklady Earth Sciences. 2017;**475**(2):939-941. DOI: 10.1134/S1028334X17080190

[5] Grigoryeva NY, Chistyakova LV, Liss AA. Spectroscopic techniques for estimation of physiological state of blue-green algae after weak external action. Oceanology. 2018;**58**(6):896-904. DOI: 10.1134/S0001437018060061

[6] Jhangirov TR, Perkov AS, Grigoryeva NY, Chistyakova LV, Liss AA. Application of linear discriminant analysis for classification of cyanobacteria by intrinsic fluorescence spectra. Izvestiya SPbGETU "LETI". 2018;**5**:45-55 (in Russian)

[7] Grigoryeva NY, Chistyakova LV, Liss AA, Klionskiy DM, Perkov AS, Zhangirov TR. Neural networks application for automation of environmental monitoring of cyanobacterial blooms in open water. In: Proceedings of XXI IEEE International Conference, Soft Computing and Measurements (SCM), 23-25 May 2018, Saint-Petersburg, Russia. Vol. 2. Saint-Petersburg: Izd. SPbGETU "LETI"; 2018. pp. 210-213 (in Russian) ISBN: 978-5-7629-2239-5

[8] Papageorgiou GC. Fluorescence emission from the photosynthetic apparatus. In: Eaton-Rye J, Tripathy B, Sharkey T, editors. Photosynthesis. Advances in Photosynthesis and Respiration. Vol. 34. Dordrecht: Springer; 2012. pp. 415-443. DOI: 10.1007/978-94-007-1579-0_18

[9] Yokono M, Takabayashi A, Akimoto S, Tanaka A. A megacomplex composed of both photosystem reaction centres in higher plants. Nature Communications. 2015;**6**:6675. DOI: 10.1038/ncomms7675

[10] Abed RM, Dobretsov S, Sudesh K. Applications of cyanobacteria in biotechnology. Journal of Applied Microbiology. 2009;**106**(1):1-12. DOI: 10.1111/j.1365-2672.2008.03918.x

[11] Vijayakumar S, Menakha M. Pharmaceutical applications of cyanobacteria–A review. Journal of Acute Medicine. 2015;**5**(1):15-23. DOI: 10.1016/j.jacme.2015.02.004

[12] Singh S, Kate BN, Banerjee UC. Bioactive compounds from cyanobacteria and microalgae: An overview. Critical Reviews in Biotechnology. 2005;**25**(3):73-95. DOI: 10.1080/07388550500248498

[13] Liu L. New bioactive secondary metabolites from cyanobacteria [thesis]. Helsinki: University of Helsinki; 2014

[14] Grewe CB, Pulz O. The biotechnology of cyanobacteria. In: Ecology of Cyanobacteria II. Netherlands: Springer; 2012. pp. 707-739. DOI: 10.1007/978-94-007-3855-3_26

[15] Rastogi RP, Sinha RP. Biotechnological and industrial significance of cyanobacterial secondary metabolites. Biotechnology Advances.

2009;**27**(4):521-539. DOI: 10.1016/j.biotechadv.2009.04.009

[16] Anahas AM, Muralitharan G. Characterization of heterocystous cyanobacterial strains for biodiesel production based on fatty acid content analysis and hydrocarbon production. Energy Conversion and Management. 2018;**157**:423-437. DOI: 10.1016/j.enconman.2017.12.012

[17] Lei Y, Chen W, Mulchandani A. Microbial biosensors. Analytica Chimica Acta. 2006;**568**(1-2):200-210. DOI: 10.1016/j.aca.2005.11.065

[18] Reshetilov AN. Biosensor development in Russia. Biotechnology Journal: Healthcare Nutrition Technology. 2007;**2**(7):849-862. DOI: 10.1002/biot.200700021

[19] Ignatov SG, Ferguson JA, Walt DR. A fiber-optic lactate sensor based on bacterial cytoplasmic membranes. Biosensors and Bioelectronics. 2001;**16**(1-2):109-113. DOI: 10.1016/S0956-5663(00)00144-5

[20] Gault PM, Marler HJ, editors. Handbook on Cyanobacteria: Biochemistry, Biotechnology and Applications, Bacteriology Research Developments Series. New York: Nova Science Publishers; 2009. 538 p. ISBN: 978-1-60741-092-8

[21] Whitton BA editor. Ecology of Cyanobacteria II: Their Diversity in Space and Time. Netherlands: Springer Science & Business Media; 2012. 760 p. ISBN: 978-9-40073-855-3

[22] Seckbach J, editor. Algae and Cyanobacteria in Extreme Environments. Israel: Springer Science & Business Media; 2007. 786 p. DOI: 10.1007/978-1-4020-6112-7

[23] Huisman J, Matthijs HC, Visser PM, editors. Harmful Cyanobacteria, Aquatic Ecology Series. Vol. 3. Dordrecht: Springer; 2005. 243 p. DOI: 10.1007/1-4020-3022-3

[24] Bryant DA, editor. The Molecular Biology of Cyanobacteria, Advances in Photosynthesis. Vol. 1. Dordrecht: Kluwer Academic Press; 1994. 881 p. DOI: 10.1007/978-94-011-0227-8

[25] Green BR, Parson WW, editors. Light-Harvesting Antennas in Photosynthesis, Advances in Photosynthesis and Respiration. Vol. 13. Dordrecht: Springer Netherlands Kluwer Academic Publishers; 2003. 514 p. DOI: 10.1007/978-94-017-2087-8

[26] Blankenship RE. Molecular Mechanisms of Photosynthesis. 2nd ed. John Wiley & Sons; 2014. 314 p. DOI: 10.1002/9780470758472

[27] Granéli E, Turner JT, editors. Ecology of Harmful Algae. Berlin: Springer Science & Business Media; 2006. 406 p. DOI: 10.1007/978-3-540-32210-8

[28] Suggett DJ, Borowitzka MA, Prášil O, editors. Chlorophyll a Fluorescence in Aquatic Sciences: Methods and Applications, Developments in Applied Phycology. Vol. 4. Dordrecht, The Netherlands: Springer; 2010. 326 p. DOI: 10.1007/978-90-481-9268-7

[29] Ke B. Photosynthesis: Photobiochemistry and photobiophysics. In: Govindjee S, editor. Advances in Photosynthesis. Vol. 10. Dordrecht: Kluwer Academic Publishers; 2001. 763 p. DOI: 10.1007/0-306-48136-7

[30] Staehelin LA, Arntzen CJ, editors. Photosynthesis III: Photosynthetic Membranes and Light Harvesting Systems, Encyclopedia of Plant Physiology. Vol. 19. Berlin: Springer; 1986. 802 p. DOI: 10.1007/978-3-642-70936-4

[31] Whitton BA. Diversity, ecology, and taxonomy of the cyanobacteria. In: Mann NH, Carr NG, editors. Photosynthetic Prokaryotes. New York: Plenum Press; 1992. pp. 1-51. DOI: 10.1007/978-1-4757-1332-9_1

[32] Wehrmeyer W. Phycobilisomes: Structure and function. In: Wiessner W, Robinson DG, Starr RC, editors. Cell Walls and Surfaces, Reproduction, Photosynthesis. Experimental Phycology, I. Berlin:Springer; 1990. p. 158-172. DOI:10.1007/978-3-642-48652-4_12

[33] Gantt E. Phycobilisomes. Annual Review of Plant Physiology. 1981;**32**(1):327-347. DOI: 10.1146/annurev.pp.32.060181.001551

[34] Glazer AN. Phycobilisome—A macromolecular complex optimised for light energy transfer. Biochimica et Biophysica Acta. 1984;**768**:29-51. DOI: 10.1016/0304-4173(84)90006-5

[35] MacColl R. Cyanobacterial phycobilisomes. Journal of Structural Biology. 1998;**124**(2-3):311-334. DOI: 10.1006/jsbi.1998.4062

[36] Förster T. Delocalized excitation and excitation transfer. In: Sinanoglu O, editor. Modern Quantum Chemistry Istanbul Lectures. Vol. 3. New York: Academic Press; 1965. pp. 93-137

[37] Şener M, Strümpfer J, Hsin J, Chandler D, Scheuring S, Hunter CN, et al. Förster energy transfer theory as reflected in the structures of photosynthetic light-harvesting systems. Chemphyschem. 2011;**12**(3):518-531. DOI: 10.1002/cphc.201000944

[38] Sauer K. Primary events and the trapping of energy. In: Govindjee S, editor. Bioenergetics of Photosynthesis. London, New York: Academic Press; 1975. pp. 115-181

[39] Mullineaux CW. Phycobilisome-reaction centre interaction in cyanobacteria. Photosynthesis Research. 2008;**95**(2-3):175-182. DOI: 10.1007/s11120-007-9249-y

[40] Wolf E, Schussler A. Phycobiliprotein fluorescence of Nostoc punctiforme changes during the life cycle and chromatic adaptation: Characterization by spectral confocal laser scanning microscopy and spectral unmixing. Plant, Cell and Environment. 2005;**28**(4):480-491. DOI: 10.1111/j.1365-3040.2005.01290.x

[41] Murton J, Nagarajan A, Nguyen AY, Liberton M, Hancock HA, Pakrasi HB, et al. Population-level coordination of pigment response in individual cyanobacterial cells under altered nitrogen levels. Photosynthesis Research. 2017;**134**(2):165-174. DOI: 10.1007/s11120-017-0422-7

[42] Vermaas WF, Timlin JA, Jones HD, Sinclair MB, Nieman LT, Hamad SW, et al. In vivo hyperspectral confocal fluorescence imaging to determine pigment localization and distribution in cyanobacterial cells. Proceedings of the National Academy of Sciences. 2008;**105**(10):4050-4055. DOI: 10.1073/pnas.0708090105

[43] Majumder EL, Wolf BM, Liu H, Berg RH, Timlin JA, Chen M, et al. Subcellular pigment distribution is altered under far-red light acclimation in cyanobacteria that contain chlorophyll f. Photosynthesis Research. 2017;**134**(2):183-192. DOI: 10.1007/s11120-017-0428-1

[44] Nozue S, Katayama M, Terazima M, Kumazaki S. Comparative study of thylakoid membranes in terminal heterocysts and vegetative cells from two cyanobacteria, Rivularia M-261 and *Anabaena variabilis*, by fluorescence and absorption spectral microscopy. Biochimica et Biophysica Acta

(BBA)-Bioenergetics. 2017;**1858**(9):742-749. DOI: 10.1016/j.bbabio.2017.05.007

[45] Ying L, Huang X, Huang B, Xie J, Zhao J, Sheng Zhao X. Fluorescence emission and absorption spectra of single Anabaena sp. strain PCC7120 cells. Photochemistry and Photobiology. 2002;**76**(3):310-313. DOI:10.1562/0031-8655(2002)0760310feaaso2.0.co2

[46] Kumazaki S, Hasegawa M, Ghoneim M, Shimizu Y, Okamoto K, Nishiyama M, et al. A line-scanning semi-confocal multi-photon fluorescence microscope with a simultaneous broadband spectral acquisition and its application to the study of the thylakoid membrane of a cyanobacterium Anabaena PCC7120. Journal of Microscopy. 2007;**228**(2):240-254. DOI: 10.1111/j.1365-2818.2007.01835.x

[47] Millach L, Obiol A, SolÉ A. Esteve I. A novel method to analyse in vivo the physiological state and cell viability of phototrophic microorganisms by confocal laser scanning microscopy using a dual laser. Journal of Microscopy. 2017;**268**(1):53-65. DOI: 10.1111/jmi.12586

[48] Mullineaux CW. FRAP analysis of photosynthetic membranes. Journal of Experimental Botany. 2004;**55**(400):1207-1211. DOI: 10.1093/jxb/erh106

[49] Mullineaux CW, Tobin MJ, Jones GR. Mobility of photosynthetic complexes in thylakoid membranes. Nature. 1997;**390**(6658):421-424. DOI: 10.1038/37157

[50] Liu LN, Aartsma TJ, Thomas JC, Zhou BC, Zhang YZ. FRAP analysis on red alga reveals the fluorescence recovery is ascribed to intrinsic photoprocesses of phycobilisomes than large-scale diffusion. PLoS One. 2009;**4**(4):e5295. DOI: 10.1371/journal.pone.0005295

[51] Vitali M, Reis M, Friedrich T, Eckert HJ. A wide-field multi-parameter FLIM and FRAP setup to investigate the fluorescence emission of individual living cyanobacteria. In: Proceedings of SPIE, Laser Applications in Life Sciences, 9-10 June 2010, Oulu, Finland. Vol. 7376. Finland: International Society for Optics and Photonics; 2010. pp. 737610-737610-6. DOI: 10.1117/12.871520

[52] Yang S, Su Z, Li H, Feng J, Xie J, Xia A, et al. Demonstration of phycobilisome mobility by the time-and space-correlated fluorescence imaging of a cyanobacterial cell. Biochimica et Biophysica Acta (BBA)—Bioenergetics. 2007;**1767**(1):15-21. DOI: 10.1016/j.bbabio.2006.11.012

[53] Blackwell M, Gibas C, Gygax S, Roman D, Wagner B. The plastoquinone diffusion coefficient in chloroplasts and its mechanistic implications. Biochimica et Biophysica Acta (BBA)—Bioenergetics. 1994;**1183**(3):533-543. DOI: 10.1016/0005-2728(94)90081-7

[54] Drepper F, Carlberg I, Andersson B, Haehnel W. Lateral diffusion of an integral membrane protein: Monte Carlo analysis of the migration of phosphorylated light-harvesting complex II in the thylakoid membrane. Biochemistry. 1993;**32**(44):11915-11922. DOI: 10.1021/bi00095a022

[55] Borlinghaus R. Colours count: How the challenge of fluorescence was solved in confocal microscopy. Modern research and educational topics in microscopy. In: Méndez-Vilas A, Díaz J, editors. Colours Count: How the Challenge of Fluorescence was Solved in Confocal Microscopy. Spain: FORMATEX; 2007. pp. 890-899. DOI:10.1.1.564.5122

[56] Li D, Shao L, Chen BC, Zhang X, Zhang M, Moses B, et al. Extended-resolution structured illumination imaging of endocytic and cytoskeletal

dynamics. Science. 2015;**349**:6251, aab3500. DOI: 10.1126/science.aab350

[57] van de Linde S, Heilemann M, Sauer M. Live-cell super-resolution imaging with synthetic fluorophores. Annual Review of Physical Chemistry. 2012;**63**:519-540. DOI: 10.1146/annurev-physchem-032811-112012

Chapter 4

From Sophisticated Analysis to Colorimetric Determination: Smartphone Spectrometers and Colorimetry

Volkan Kılıç, Nesrin Horzum and Mehmet Ertugrul Solmaz

Abstract

Smartphone-based spectrometer and colorimetry have been gaining relevance due to the widespread advances of devices with increasing computational power, their relatively low cost and portable designs with user-friendly interfaces, and their compatibility with data acquisition and processing for "lab-on-a-chip" systems. They find applications in interdisciplinary fields, including but not limited to medical science, water monitoring, agriculture, and chemical and biological sensing. However, spectrometer and colorimetry designs are challenging tasks in real-life scenarios as several distinctive issues influence the quantitative evaluation process, such as ambient light conditions and device independence. Several approaches have been proposed to overcome the aforementioned challenges and to enhance the performance of smartphone-based colorimetric analysis. This chapter aims at providing researchers with a state-of-the-art overview of smartphone-based spectrometer and colorimetry, which includes hardware designs with 3D printers and sensors and software designs with image processing algorithms and smartphone applications. In addition, assay preparation to mimic the real-life testing environments and performance metrics for quantitative evaluation of proposed designs are presented with the list of new and future trends in this field.

Keywords: color sensing, smartphone spectrometer, image processing, colorimetry, Android application

1. Introduction

Smartphone-based analysis has recently emerged as a useful tool, and it has been found to be promising in several fields including point-of-care analyses [1], chemical and biological sensing [2], microscopy and healthcare diagnostics [3], water quality sensing for environment [4–6], leaf color analysis for agriculture [7], pH [8, 9] and glucose [10] sensing, fluorescent imaging [11], imaging cytometry [12], electrochemical sensing [13], and immunoassays [14].

With recent advances on camera and sensor technologies, current smartphones are equipped with a low-power high-performance processor with up to 2.5 GHz operating frequencies; built-in high-resolution digital camera, generally above 5 Mpixels and up to 40 Mpixels; and built-in single or dual LEDs, which allow capturing an image even in low-light conditions [15]. Moreover, they are provided with advanced onboard sensors such as moisture sensors, proximity sensors, electromagnetic compasses, accelerometers, and gyroscopes [16]. These sensors generally provide fast response, being portable and of low cost, and the ability to be used in the field without extensive training [17]. Therefore, smartphones have become a tool as powerful as low-cost computers, which lead them to be valuable instruments in the analysis.

Over the last decade, smartphones have been increasingly used in a variety of scientific fields as spectrometers [18–22] and colorimeters [23–25]. Smartphone spectrometers use the wavelength components, which give spectral information, of the collimated light from the optical source which is dispersed after interaction with samples [18]. The color spectrum image is transformed into various color spaces for the extraction of quantitative data. The wavelength of the spectrum generally changes between 400 and 700 nm because of the optical filters set in front of the camera in the manufacturing process. Spectral information has been used in many applications including water monitoring [6], gas detection [26], and food quality control [27]. Smartphone colorimeters are commonly used instruments that quantify the concentration of the samples based on color changes due to concentration (like peroxide amount [28]) or time (methylene blue degradation [6]). In the colorimetric analysis, features of the referenced images need to be extracted for training the system, mostly created with machine learning or neural networks, which perform the quantitative analysis for the test images. Smartphone colorimeters have applications in both solid samples like paper-based test and liquid samples like colored solutions [29]. Both smartphone spectrometers and colorimeters are powerful tools for rapid qualitative and quantitative analyses due to the fact that they can be used in conditions where sophisticated tools or time-consuming steps cannot be used. They are rapid and low-cost tools that require less sample consumption and provide portability to perform the analysis in remote locations or locations with poor infrastructure [17].

While smartphones offer an attractive alternative to sophisticated tools for imaging and analysis in the field, there are some concerns about their suitability for quantitative analysis [1]. First, unlike scientific cameras, smartphone cameras have mostly limited control of camera parameters like exposure time, shutter speed, ISO, and color balance, and no access to raw image data which has a linear relationship with scene radiation. In addition, image processing algorithms, such as demosaicing, noise reduction, edge sharpening, white balance, and image compression, are applied automatically and vary significantly across smartphones. These methods corrupt the linearity of the pixel intensity values which causes loss of information that can be used in quantitative analysis. So, it is difficult to set and maintain imaging parameters to get accurate and repeatable information for analysis [30]. Second, ambient light conditions are hard to control during imaging in uncontrolled environments. Last, small color changes cannot be detected in an analysis as the red, green, and blue (RGB) intensity values may not be sufficient. These aforementioned concerns make smartphones questionable for quantitative analysis. Early studies in smartphone-based colorimetric analysis for medical and scientific applications pointed out these concerns and concluded that smartphones are incapable of pathology [31] or limited due to their image quality [32].

However, subsequent researchers have proposed a smartphone camera-based microscope which captures qualitatively relevant features of malaria and tuberculosis [33]. These counter conclusions and also advancements in camera and sensors together with the increasing capability in computer processing prove that smartphones are more portable, cost-efficient, and user-friendly platforms which make them alternatives to sophisticated and high-cost devices in quantitative analysis.

For instance, it was shown that a smartphone by itself is capable of quantifying colorimetric test strips without any external attachments [1, 34]. In [35], water monitoring was implemented with a simple holding attachment. With the help of 3D printing, housing and optical components were integrated into a colorimetric plate reader for enzyme-linked immunosorbent assay (ELISA) [36]. A 3D printed custom cradle including various optical components and an external broadband source, was demonstrated in [37] for biomarker absorption analysis. To detect miRNA sequences in a liquid-based assay, a smartphone-based fluorimeter system was designed using an external laser as the source [21]. Due to precise light confinement and flexible nature, optical fibers were integrated to smartphone spectrometer and applied for food quality monitoring [18]. To have more portable and cheap spectrometers, new designs were proposed without external electrical and optical components. A fiber optical bench was assembled on top of the smartphone without external LEDs to design a surface plasmon resonance-based refractive index sensor [38]. A spectrometer system was reported in [10] to detect glucose and troponin-I using built-in flash as a light source and a compact disk for the reflection grating.

In the colorimetric analysis, color information could be obtained with paper-based sensors to quantify the color variation in different color spaces such as RGB, HSV, and L*a*b* [24, 39–41]. In [25], alcohol concentration in saliva was detected using paper-based test converting images from RGB to HSV color space. This was the conventional approach which needs only a smartphone camera to capture and process an image. Besides the conventional approach, non-conventional approaches were applied on liquid samples in vials for detection of chlorine in water [42], and ripeness estimation of fruits [7]. Here, quantification was calculated using analytical formulas extracted from color space parameters. However, it is prone to deviation due to the disadvantages of JPEG images such as low bit depth and heavy post-processing (white balance, contrast, and brightness adjustment) [43, 44]. On the other hand, it was shown in [28, 41] that JPEG images could be used in the colorimetric analysis when advanced algorithms like machine learning were used to process the images. Unfortunately, advanced algorithms need more computational power as they need much larger datasets for training and testing the images. To address this issue, it was reported to use local database referenced with a single image for the quantification of the concentration level of solutions [45]. The proposed design was applied on nitrite, phosphate, chromium, and phenol solutions to quantify the concentration value using a single reference image which was captured and processed initially.

In literature, there are many lab-on-a-chip designs proposed for various applications [17, 46]. Bisphenol-A (BPA) detection in distilled and commercial water samples was demonstrated in [29] where a plastic fiber-based smartphone spectrometer with a custom-designed immersion probe and a cradle were used. The smartphone spectrometer was converted to a reflection probe spectrometer working within the visible spectrum. Explosive types were detected in [17] with the paper-based test using hierarchical clustering analysis and principal component analysis (PCA) regarding the color discrimination of the explosives. A closed chamber was used to eliminate ambient light conditions during imaging. Linear correlation and PCA methods were employed, respectively, for univariate and multivariate analyses [46]. It was reported that univariate analysis did not give statistically significant results due to ambient light conditions as imaging was not performed in a controlled environment. However, multivariate analysis gave promising results running PCA methods on eight different color spaces and clustering methods including red, green, blue, hue, saturation, value, lightness, and intensity parameters. A smartphone-based colorimetric reader for ELISA was proposed in [47] where imaging was performed in a controlled environment illuminated from the bottom. Ozcan and his research group [12, 14, 48–51] had many designs for specific applications such as food allergen testing [48], urine [14] and blood [12] analysis, immunoassays [49], microscopy [50], and cytometry [51].

The rest of this chapter is organized as follows. The next section introduces the hardware designs for smartphone-based spectrometer and colorimetry. Section 3 presents the mobile apps and image processing algorithms to be used in smartphones. Section 4 describes assay preparation to be used in testing the performance of both spectrometer and colorimeter. Section 5 details the metric to evaluate the performance of the proposed designs. Closing remarks are given in Section 6.

2. Hardware designs

In this section, a brief review of hardware designs of smartphone spectrometer and colorimetry is presented. Because of space constraints, only two examples of spectrometer and colorimetry are demonstrated in Section 2.1 and 2.2, respectively.

2.1 Spectrometer

Recently, there has been a growing variety of spectrometer designs for specific applications. In [19], a compact imaging spectrometer was reported equipped with motorized selfie stick for remote sensing. Another spectrometer was designed to calculate spectra and quantify analytes by the assembly of medium-density-fiberboard, a DVD slice for diffraction grid, and mini incandescent lamps [20]. A flexible fiber bundle probe was integrated with a custom-designed cradle to convey spectra data to smartphone camera for food quality monitoring [18].

The next design [6] is illustrated in **Figure 1a**, which is a low-cost, portable, plastic fiber-based spectrometric smartphone to analyze dye adsorption for field-deployable environmental and wastewater management. In the design, the rear camera of LG G4 (1/2.6″ sensor size with 5312 × 2988 resolution, 1.12 μm pixel size) was used to collect the spectral data of the assays.

The smartphone was fitted into a custom-designed cradle assembled with hot-plug apparatus toting a diffraction grating, and the whole part was connected with smartphone case and cuvette holder manufactured from Acrylonitrile Butadiene

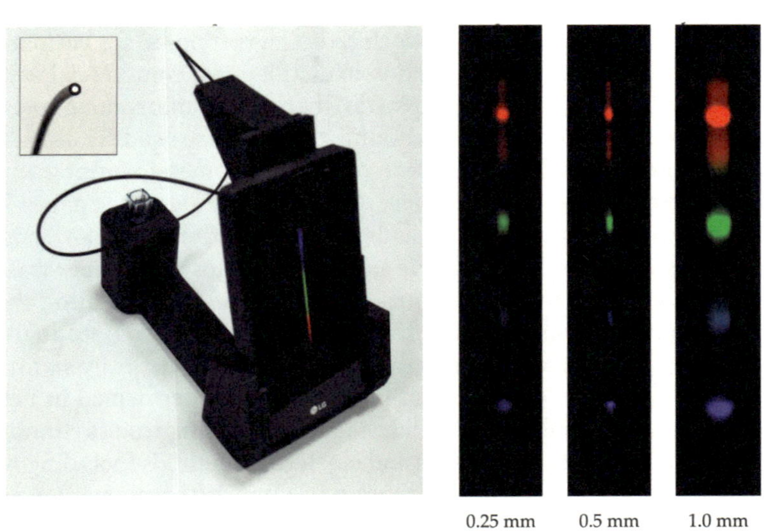

Figure 1.
(a) The smartphone spectrometer with the inset (top-left) of a plastic fiber assembled into the built-in flash.
(b) The spectral images from different-sized plastic fibers are shown.

Styrene polymer using a 3D printer (Zortrax M200) with a 150-g polymer. Two pieces of plastic fiber cables were used in the design. The first 1.5-mm-diameter fiber carried the light from the smartphone flash to the cuvette while the second fiber cable with a diameter of 0.25 mm transmitted the light from cuvette to the camera which passed through the assay. The diameter of the second cable was critical as the light for spectral data was carried with this cable. Therefore, the effect of diameter on spectral data was analyzed using 0.25-, 0.5-, and 1.0-mm cables as given in **Figure 1b**, and 0.25-mm diameter was found to be adequate based on this experiment. A custom cradle was specially designed to align plastic optical fibers with smartphone optical components. As the cradle was solid, the solution could be placed into the cuvette slot. In order to simplify the spectrometer system design, no collector lens or mirrored components were placed in the light path. Besides cost, the most important factor in choosing plastic optical fibers instead of glass-based fiber optics was the ability to use plastic optical fibers without special tools for stripping and cutting.

To test the performance of the system, methylene blue (MB) solutions were prepared with different amounts and their respective spectral views are given in **Figure 2**. At the top row, the concentration of the solution varies from 0 ppm (most left), which

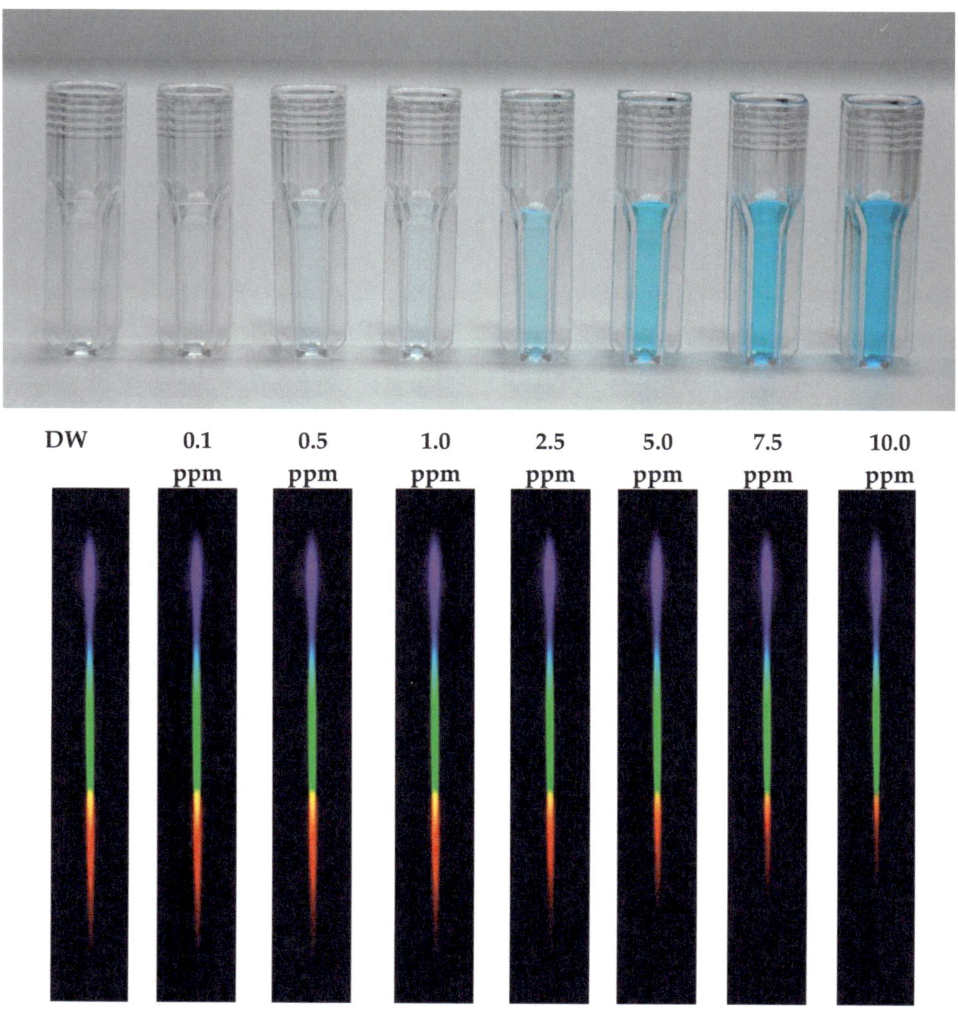

Figure 2.
MB solutions with a spectral view from 0.1 to 10 ppm.

corresponds to distilled water (DW), to 10 ppm (most right). The spectrum views are given at the bottom row where the reduction in red intensity with the concentration is quite visible.

This design was further improved to make it compatible with immersion probe, which made it more practical and user-friendly as illustrated in **Figure 3** [29]. Schematic diagram of the 3D printed cradle is described in **Figure 3a**. The immersion probe was attached to the fiber-coupled smartphone flashlight and the reflection caused by radiation is carried to the camera via the grating. Plastic (PMMA)-based bifurcated fiber bundle was used to manufacture the probe with the diameter of 0.5 mm (also known as Y-cable) as shown in **Figure 3b**. The overall design is illustrated in **Figure 3c**. It was reported that no additional optical components were used in the reflection-based smartphone spectrometer system. The spectrum views

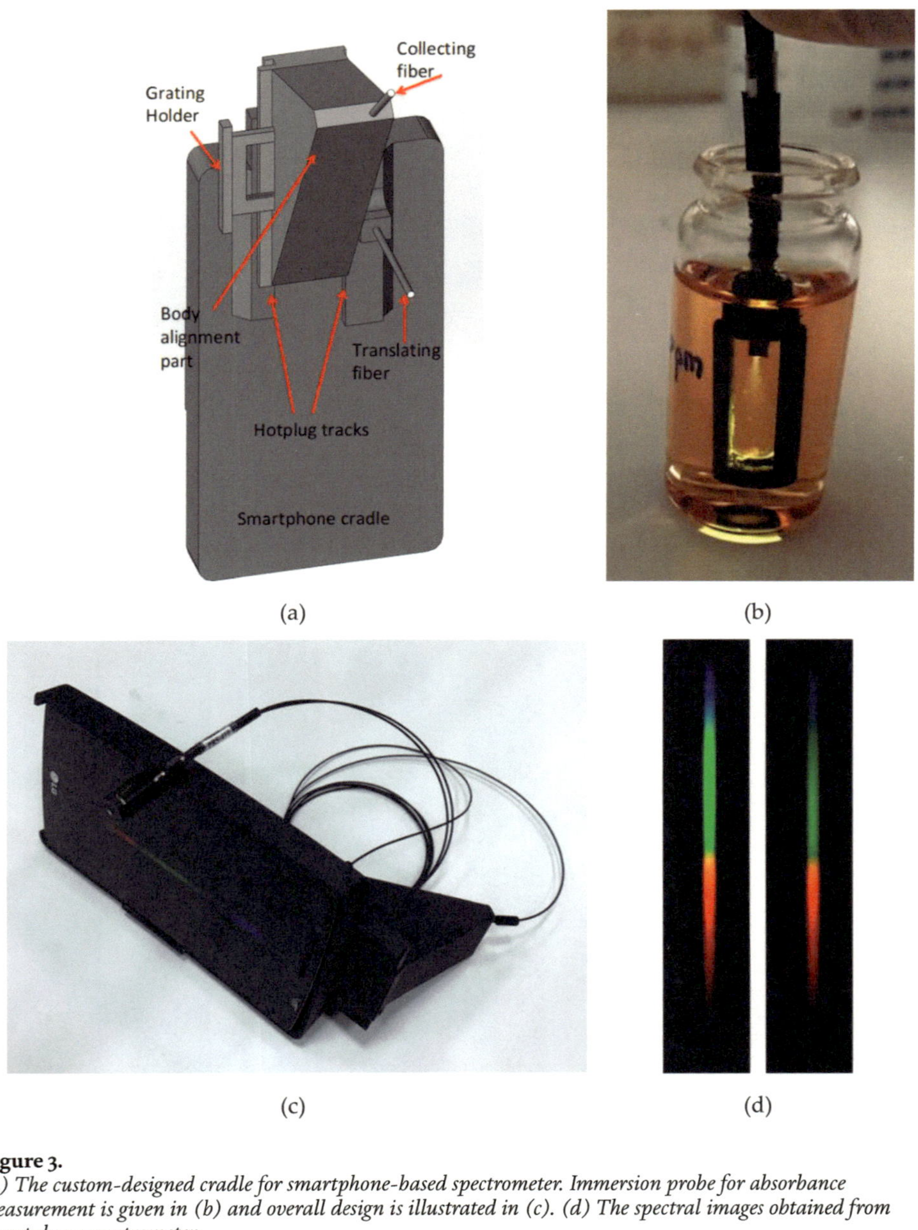

(a) (b)

(c) (d)

Figure 3.
(a) The custom-designed cradle for smartphone-based spectrometer. Immersion probe for absorbance measurement is given in (b) and overall design is illustrated in (c). (d) The spectral images obtained from smartphone spectrometer.

of blank solution (left) and 5-ppm BPA solution (right) are given in **Figure 3d** to demonstrate color variation can be detected with the spectrometer system.

2.2 Colorimetry

As an alternative to spectrometric analysis, colorimetry is also widely used in many applications including food allergen testing [48], albumin testing in urine analysis [14], blood analysis [12], pH quantification [41], and water monitoring [45].

A digital tube reader designed in the 3D printer was equipped with two interchangeable LEDs to illuminate the test and control tubes so that the absorption spectrum of the colorimetric assay could be analyzed [48]. An albumin tester platform was proposed in [14] using an optomechanical attachment aligned with a smartphone camera. The 3D printed cradle was integrated to a compact laser diode, two AA batteries, a plastic lens, and an emission interference filter. An albumin-based fluorescent signal was obtained from the test tube by a digital fluorescent tube reader to calculate the albumin concentration values after comparison with a control tube. In [12], blood analysis was implemented with an integration of red blood cell counting, white blood cell counting, and hemoglobin measurement devices to smartphone cradle.

Smartphone-based colorimetric detection of pH, which varies between 0 and 14.0, was investigated with paper-based test [41]. The performance of the system was tested under two conditions: controlled and ambient illumination environments. To create controlled illumination settings, 3D printed cradle was equipped with apparatus which eliminates the interference of the present light as shown in **Figure 4a**. Four strips of same pH level were located side by side for imaging with an apparatus, then color calibration and white balancing were performed for those strips with the X-Rite ColorChecker Passport. The imaging was continued with replacing the strips in six different orientations as in **Figure 4b**.

Later, random orientations as shown in **Figure 4c** were used to mimic scenarios that could happen when untrained users take a picture. The reason for using a group of four strips is to see the effect of luminance variation due to their positioning with respect to the camera flash.

For the ambient illumination environments, no apparatus was used, and instead of using the smartphone flash as a light source, sunlight and fluorescent and halogen sources were used. To test the system under challenging conditions, the light sources were used in solo and in dual and triple combinations. The images captured in both controlled and ambient illumination environments with a different

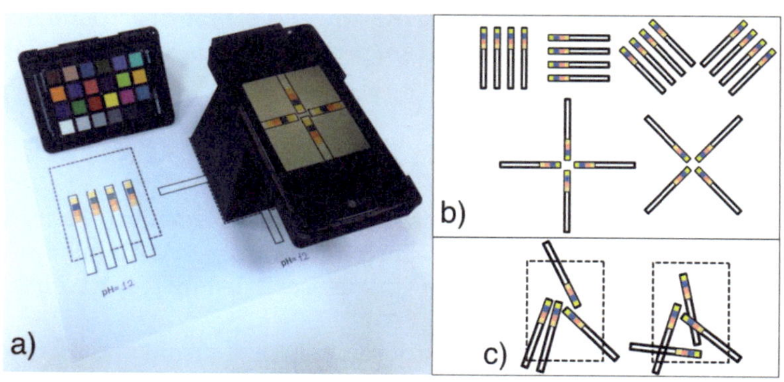

Figure 4.
The overall smartphone-based colorimetry with apparatus and X-rite ColorChecker passport for color correction are shown in (a). The pH strips with various orientations used in imaging are given in (b), and (c) shows random orientations and positions of the test strips inside the smartphone field of view for dual-illumination tests.

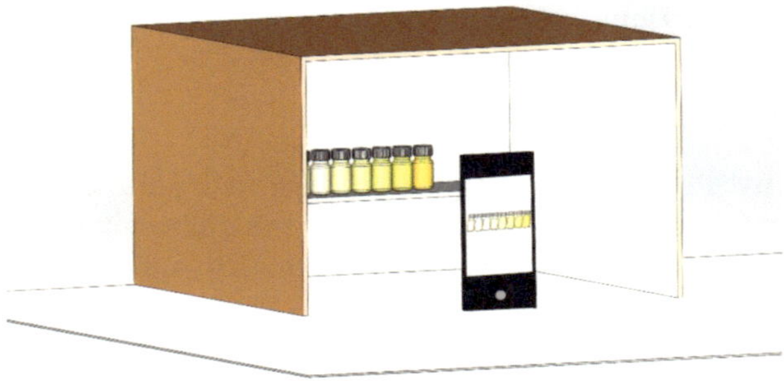

Figure 5.
Experimental setup proposed in [45].

replacement of the strips were used to train the machine learning algorithm which was designed to quantify the pH values accurately. Since the training set was enriched with the images captured in various and complicated scenarios, it was reported that pH values were detected with 100% accuracy.

One possible drawback of the colorimetry method in [41] is its computational cost due to its large training dataset. To address this issue, single-image referenced colorimeter was proposed in [45]. The system was simplified in the sense of both hardware and software design. Instead of using machine learning algorithm which needs a large dataset, it used local dataset created by a user with a single reference image. In addition, images from the local dataset were compared with test images using color-matching algorithms computationally cheaper than machine learning algorithms. The hardware design was also simplified into a cardboard box as shown in **Figure 5**. It was painted white, and white light-emitting diodes were mounted to the box ceiling to minimize the ambient illumination effects. A holder platform with the same height as the camera was placed for the assays to maintain the same distance for imaging. The system was tested on four different (nitrite, phosphate, chromium, and phenol) assays, and it was reported that the performance accuracy was between 76 and 100% depending on the assay types.

3. Software designs

The previous section described the hardware designs for smartphone-based spectrometry and colorimetry. This section presents mobile applications and algorithms proposed for these designs.

3.1 Mobile apps

Software applications are necessary tools due to complementary characteristics for the hardware designs of spectrometer and colorimetry. Mobile apps were therefore developed to make the overall system user-friendly [14, 15, 25, 28, 36, 45, 46, 52].

Albumin Tester [14] application was developed for Android phones to let the user determine the albumin concentration in the urine sample. To test alcohol in saliva, SPAQ [15, 25] application was developed which estimated the alcohol level based on the histogram distribution. Colorimetric Plate Reader app [36] was proposed for qualitative and quantitative ELISA test. PhotoMetrix [46] application was introduced, which runs the univariate and multivariate analyses to quantify the analytes in the samples. Colorimetric Test Reader app was presented in [52], which determines pH, protein, and glucose values in the assay.

Solmaz et al. [28] developed ChemTrainer app which quantifies the peroxide content running machine learning algorithm on the remote server as shown in **Figure 6a**. After capturing a photograph of colorimetric test strips, the app sends mean RGB values for a region of interest to the remoter server as machine learning algorithm needs only mean values for the classification. A message queue service was employed to enable multiple users to reach the server simultaneously.

Screenshots of the ChemTrainer are given in **Figure 6b**. In the opening page, there are two options for the user: either capturing a new image with "Experiment" button or using existing image from the phone with the "Load From Gallery" button. After an image is captured or loaded from the gallery, the user may proceed or retake a new image. Next, a region of interest needs to be cropped with adjustable crop box. The app calculates the average red, green, and blue values of the cropped image and sends to the server, which runs a classification algorithm that decides the class of image. In the meantime, the app displays a progress animation until the result comes back from the server.

The ChemTrainer app was further improved to be able to work with single-image reference (SIR) and named as ChemTrainerSIR [45]. In addition, it gained additional features like saving location and time data for the previous experiments, which help the user to analyze the past results when needed. The ChemTrainerSIR app is described in **Figure 7** with screenshots. There are "train" and "experiment" options on the home page as shown in **Figure 7a**. Training steps are introduced in **Figure 7** from b to g (top row), while testing steps are described at the bottom row from h to m. If the user selects the train option, some initial information needs to be entered such as the name of the chemical compound as the name of the model (e.g., phosphate), the units of measurement (e.g., ppm), and the number of samples with known concentration levels. The user either captures an image of samples or loads from the gallery. Then, the user enters the reference values for each sample which will be used later in the testing phase. All these information are stored in a designated folder in internal storage. In the testing phase, the user first selects the model, which was used in training. Then, a new image is taken to quantify the concentration value based on comparison with the reference model.

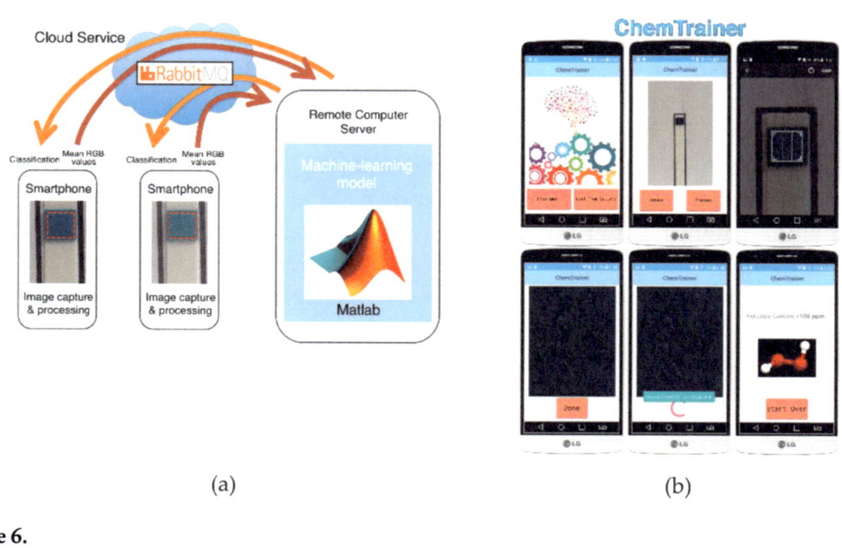

(a) (b)

Figure 6.
The communication between the smartphone and a remote server is illustrated in (a) and the developed ChemTrainer app is presented in (b).

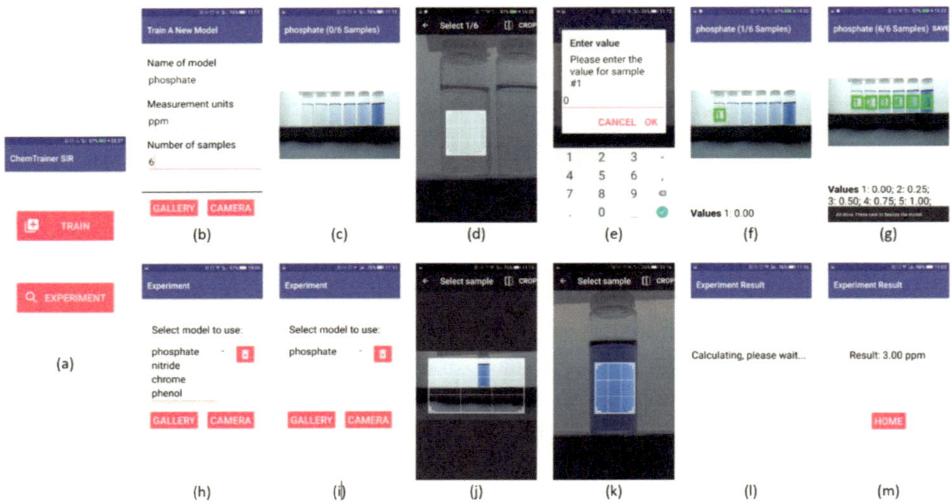

Figure 7.
Opening page of the ChemTrainerSIR app is given in (a). The top row (b–g) shows training steps, while the bottom row (h–m) presents the implementation of the app on the sample.

3.2 Image acquisition and processing

A digital image can be acquired either as monochrome (black and white) or color image using electronic equipment utilizing charge-coupled device (CCD) or complementary metal oxide semiconductor (CMOS) sensors.

These sensors use a two-dimensional array of millions of tiny light pixels to capture an image. These pixels collect photons and store them as an electrical signal after the shutter button of the camera is pressed which leads to the beginning of the exposure. The pixels are closed after the exposure finishes, and intensity value in the pixel is quantified as digital values by measuring the strength of the electrical signal, which is directly related with the number of photons stored in the pixel. However, this approach would only create a monochrome or gray scale image as the pixels are unable to distinguish photons in terms of color. A color filter needs to be placed over the pixel to capture a color image. This filter allows only one of the primary colors, that is red (R), green (G), or blue (B), to pass into the pixel, so that it stores only filtered photons for the respective color. In other words, the intensity of each pixel gives single color information which leads to a RAW image. Here, each pixel has only one of R, G, or B information while all R, G, and B values need to be known for each pixel. Therefore, demosaicing is applied to determine other two missing color values by interpolating from nearby pixels where those colors are known. After demosaicing, other methods such as white-balance, gamma correction, color space correction, and compression are applied to convert the image from RAW to a common format like JPEG.

JPEG images have a small size and they can be displayable instantly. However, there are some concerns because of the methods that are applied to convert an image from RAW to JPEG. In the conversion process, the image is compressed resulting in providing a non-linear RGB color space with only 8-bit color depth [43, 53]. However, RAW images contain original image data with 10–14 bits of color information. The conversion process corrupts the linearity of the image. A linear image conserves the relation between the intensity value and the number of photons which maintains the linearity with scene radiance. This linearity is required for quantitative scientific data acquisition in many applications [6, 30, 36, 37, 54]. As a linear image, the RAW format is therefore generally chosen. The main issue is how to reach a RAW image in smartphones. Although most semi-professional and professional cameras have access to reach the images in a RAW format, it is unconventional for smartphones. With recent developments, the latest smartphones offer access to images in the RAW format [55]. The RAW images could be found in three extensions such as ".NEF" (Nikon), "CR2" (Canon), and ".DNG." The most used format is ".DNG" as it has a common open format. Currently, no app is available to process the ".DNG" images in a smartphone yet. Therefore, free DCRAW software [18] can be used to convert ".DNG" image to tagged image file format (TIFF) for easier extraction of the R, G, and B values of the image.

RGB is the most commonly used color model in image processing. However, it can be converted to other models such as hue, saturation, and value (HSV); hue, saturation, and lightness (HSL); hue, saturation, and intensity (HSI); and lightness, green-red, and blue-yellow (L*a*b*). Hue is defined with the color portion of the color model and described with a number from 0 to 360°. Saturation is defined with the amount of gray in the color, from 0 to 100%. Value, lightness, or intensity is the brightness or intensity of the color, from 0 to 100% where black is represented with 0 while 100 is the brightest.

After the image acquisition, numerous image processing methods can be employed to improve the image visualization so that better features can be extracted from the image. The feature extraction plays a critical role in some methods like PCA, convolutional neural network, and machine learning, which interpret multiple types of information contained in an image using these features [46].

The performance of these methods was investigated with RAW and JPEG image formats with different color spaces such as RGB, HSV, and L*a*b*. RAW and JPEG image formats were studied in [6, 29] after converting images from RGB to HSV. Absorbance experiments were employed based on V components of HSV and it was reported that RAW format outperforms the JPEG formats in absorbance measurements. On the other hand, [41] showed that JPEG images gave a similar performance with RAM image if least-squares support-vector machine (LS-SVM) was employed in creating the learning model. Based on this conclusion, RGB, HSV, and L*a*b* color spaces were investigated using JPEG formats for quantifying peroxide content based on machine learning classifiers [28]. JPEG images were also used in [45] where the images were converted from RGB to L*a*b* color space. Instead of machine learning algorithms, color matching algorithms such as deltaE and color correlation methods were employed due to their simplicity. It was reported that deltaE showed superior performance with L*a*b color space for colorimetric water quality detection.

4. Assay preparation

In previous sections, various hardware and software designs were introduced for smartphone-based spectrometer and colorimetry. These designs need to be tested under the conditions that users may encounter in real life. In this section, strip and assay preparations are introduced, which are commonly used for water quality and field tests.

Colorimetric detection of pH values was studied in [41], which used pH strips to test their proposed system. First, solutions were prepared by mixing deionized water with sodium hydroxide (NaOH) and nitric acid (HNO_3) to ensure the pH values in the range of 0–14.0. During the preparations, pH values were checked with a pH meter (HI 2223, Hanna Instruments, RI, USA) calibrated with standard buffers, pH 4.0 (HI 7004) and 7.0 (HI 7007) prior to using pH indicator strips (Merck, Germany). In addition, dual-illumination tests were performed with buffer solutions (4.0–9.0, Sigma-Aldrich, USA). Before imaging pH strips, they were immersed into the pH solutions for 5 s and wiped gently with tissue paper, so that light refraction caused by the liquid drops could be minimized.

Peroxide quantification with colorimetric tests was investigated in [28] and hydrogen peroxide (H_2O_2) solutions were prepared for the peroxide test strips (Quantofix Peroxide 100). First, a stock solution with 500 ppm concentration of H_2O_2 (Sigma-Aldrich) was prepared in distilled water. The stock solution was later diluted to prepare the initial concentrations such as 1, 3, 10, 30, and 100 ppm. The peroxide test strips were dipped into these solutions for 1 s, and images were taken by smartphone after they were dried on tissue paper for 5 s.

Bisphenol-A (BPA) detection with smartphone spectrometer was demonstrated in [29]. The BPA concentration was determined with absorbance measurements using an immersion probe. The phenolic compound was put into reaction with 4-Aminoantipyrine (4-AAP) (Sigma-Aldrich, >98%) and potassium ferricyanide (Carlo Erba) for colorimetric quantification. Around 200 ppm of BPA stock solution was prepared in ethanol and then test solutions ranging from 0.1 to 10.0 ppm were prepared by serial dilution from the stock solution. The pH of all solutions was set to 8.0 using 0.25 M sodium bicarbonate ($NaHCO_3$) (Sigma-Aldrich, ≤99.7%) and distilled water. To finalize the solutions, 1.5 mL of 20.8 mM 4-AAP and 1.5 mL of 83.4 mM potassium ferricyanide solutions were mixed into 12 mL of BPA solutions. The solutions were ready for the absorbance measurement after 10 min of incubation.

Single-image-referenced colorimetric water quality detection in [45] was performed using four different analyte solutions. The first solution is nitrite (NO_2^-); it reacts with sulfanilamide to form diazonium ion which was coupled with *N*-(1-naphthyl) ethylene diamine dihydrochloride (NED) to produce an azo dye. A stock solution (821 ppm) was prepared by dissolving 1.232 g sodium nitrite in 1 L of distilled water. The buffer stock (164 ppm) and standard (1.64 ppm) solutions were prepared to obtain solutions in concentrations of 0, 0.05, 0.10, 0.20, 0.40, and 0.50 ppm by dilution. The pH of solutions was adjusted with 1 N HCl or NH_4OH to give a pH between 5.0 and 9.0. For the color reagent, 10 g of sulfanilamide was dissolved in a mixture of 100 mL of 85% phosphoric acid and 800 mL of distilled water. Then, 1 g of NED was added and diluted to 1 L. After addition of this reagent (2.0 mL) to NO_2^--containing solutions (50 mL), a stable red-violet color was observed. The second solution is the phosphate (PO_4^{-3}) determined by the reaction of ammonium molybdate and antimony potassium tartrate in an acidic medium with solutions containing phosphorus to form an antimony-phosphomolybdate complex which is reduced by ascorbic acid. A stock solution of 50 ppm $PO_4^{-3} - P$ was prepared by dissolving 219.5 mg of anhydrous potassium dihydrogenphosphate in 1 L of distilled water. The solutions of PO_4^{-3} standards in concentrations of 0, 0.25, 0.50, 0.75, 0.40, 1.00, and 3.00 ppm were prepared by serial dilution. A composite reagent containing (15 mL, 0.032 M), (5 mL, 0.008 M), and (30 mL, 0.100 M) in 50 mL of 5 N sulfuric acid was provided after each reagent addition. The pH of solution was controlled by 1 drop of phenolphthalein addition into the 50 mL of sample solution. If a red color develops, a strong acid is dropped till the color disappears. Then, 8 mL of the composite reagent was added to the sample solutions and they are allowed to stand for at least 10 min to measure the stable blue color. Hexavalent chromium (Cr(VI)) detection is performed by the formation of a colored complex resulting from the reaction of Cr(VI) with diphenylcarbazide in an acidic medium. A stock solution of 500 ppm Cr(VI) was prepared by dissolving 141.4 mg of dried potassium dichromate in distilled water and diluted to 100 mL. To prepare a standard solution of 5.00 ppm Cr(VI), 1.00 mL of the stock solution was diluted to 100 mL. The solutions of Cr(VI) standards in concentrations of 0, 0.05, 0.10, 0.20, 0.40, 0.50, and 1.00 ppm were prepared by serial dilution. The pH of the standards was adjusted to 2.0 ± 0.5 with 0.25 mL of nitric acid and 0.2 N sulfuric acid. Around 2.0 mL of complexation reagent, freshly prepared by dissolving 250 mg of 1,5-diphenylcarbazide in 50 mL of acetone and stored in a dark glass bottle, was added to 100 mL of Cr(VI) solutions. After 5–10 min, a red-violet color was visible. At last, phenolic compounds react with 4-aminoantipyrine (AAP) in the presence of potassium ferricyanide at alkaline pH to form a colored antipyrine dye. A preliminary distillation is required to eliminate or minimize possible interfering materials. Copper(II) sulfate pentahydrate (5 mL, 7.6 M) solution was added to a 500-mL sample after the pH of the solution was adjusted to 1–2 with 85% phosphoric acid (v/v) (1 + 9), transferred to the distillation apparatus. The distillation was stopped when the volume of distillate became about 400 mL, and 100 mL of distilled water was added to the flask. Then, distillation was continued until 500 mL of distillate was collected. About 5 mL of ammonium chloride solution was added to a 100-mL portion of distilled sample and pH of the solution was adjusted to pH 10 ± 0.2 with ammonia solution. 4-Aminoantipyrine solution (2 mL, 0.098 M, daily prepared) and potassium hexacyanoferrate(III) solution (2 mL, 0.24 M, weekly prepared) were subsequently added with stirring.

5. Performance metrics

Quantitative performance evaluations of smartphone-based spectrometer and colorimeter are an important factor in the development of new algorithms and designs. Standard metrics for regression and classification problems can be used to assess smartphone-based system performance. The importance of metrics varies for each sensing scheme.

In a spectrometer, the absorbance spectrum needs to be calculated using multi-colored images. RGB images are mostly converted to HSV images and value (V of HSV) is used to calculate the absorbance (A) using the Beer-Lambert law [56],

$$A = \log_{10}\frac{I_0}{I} \tag{1}$$

where I_0 is the transmitted light intensity of reference solution (mostly distilled water), and I is the transmitted light intensity of the other solutions. After the absorbance graph is plotted with respect to wavelength, the reference wavelength point which gives the maximum absorbance of the reference solution is selected. Then, the calibration curve, which is basically the linear regression line, is plotted with respect to the reference wavelength point to calculate R^2 (the coefficient of determination). R^2 is the first metric to evaluate to assess the performance of the model. R^2 values greater than 0.9 are acceptable values, although a larger coefficient is accepted as a more successful result. Next, evaluation term is the limit of detection (LOD) defined as the lowest quantity or concentration of an analyte that can be reliably detected with a given analytical method. It is calculated as three standard deviations above the reference solution. The slope of the calibration curve is the sensitivity of the spectrometer.

In classification-based colorimetry, the following metrics are available: classification accuracy, sensitivity (recall), specificity, precision, and f1-score. These metrics are the same in traditional machine learning classification tasks and can be extracted from the confusion matrix. Classification accuracy is detection accuracy in the case of analytical detection. For binary classification problems with only two classes, the receiver operation characteristic (ROC) curve and area under curve (AUC) are additional metrics. In a confusion matrix, rows represent the instances in an actual (true) class while columns represent the instances in a predicted class. To calculate the detection accuracy, diagonal elements of the confusion matrix are summed and divided by the total number of data points. Precision is calculated by the ratio of true positive events to the sum of true and false positive events as given below:

$$\text{Precision} = \frac{\textit{True Positive}}{\textit{True Positive} + \textit{False Positive}} \quad (2)$$

The sensitivity (recall) is the ratio of true positive to the sum of true positive and false negative:

$$\text{Sensitivity} = \frac{\textit{True Positive}}{\textit{True Positive} + \textit{False Negative}} \quad (3)$$

Lastly, f1 score is the harmonic average of precision and the recall and is equal to 1 for perfect precision and recall:

$$\text{f1} = 2 \times \frac{\textit{Precision} \times \textit{Recall}}{\textit{Precision} + \textit{Recall}} \quad (4)$$

Regression and classification metrics should be chosen based on the colorimetric detection scheme. Spectrometric detection requires the use of regression metrics while the detection of discrete color change should be assessed with classification metrics.

6. Chapter summary

In this chapter, a review of smartphone-based colorimetric determination of chromogenic assays has been provided on color spaces, existing color matching and detection techniques, hardware and software designs, and performance metrics that have been developed over the past few decades.

After a broad survey of the smartphone spectrometers, a technical background of the methods for image acquisition system, image analysis, and measurement procedure, which are commonly used as baseline methods in the literature, was introduced with their basic mathematical, statistical concepts and definitions, which are required for understanding the mathematics and techniques behind the proposed colorimetric detection methods.

In addition, portable hardware designs compatible with smartphones and their Android applications were introduced with fundamental differences including physical setup, interfaces, and challenges.

Moreover, performance metrics were analyzed in order to see which aspects are considered more in the evaluation and impacts of these perspectives on the evaluation results.

Acknowledgements

The authors thank Dr. Gazihan Alankuş for mobile apps and the "Ekosfer Laboratory and Research Services" for their assistance with the colorimetric assays. This research was supported by the scientific research projects coordination unit of Izmir Katip Celebi University (project no. 2018-ÖDL-MÜMF-0021).

Conflict of interest

The authors declare no conflict of interest.

Author details

Volkan Kılıç[1]*, Nesrin Horzum[2] and Mehmet Ertugrul Solmaz[1]

1 Department of Electrical and Electronics Engineering, Izmir Katip Celebi University, Izmir, Turkey

2 Department of Engineering Sciences, Izmir Katip Celebi University, Izmir, Turkey

*Address all correspondence to: volkan.kilic@ikc.edu.tr

References

[1] Shen L, Hagen JA, Papautsky I. Point-of-care colorimetric detection with a smartphone. Lab on a Chip. 2012;**12**(21):4240-4243

[2] Grudpan K, Kolev SD, Lapanantnopakhun S, McKelvie ID, Wongwilai W. Applications of everyday IT and communications devices in modern analytical chemistry: A review. Talanta. 2015;**136**:84-94

[3] Contreras-Naranjo JC, Wei Q, Ozcan A. Mobile phone-based microscopy, sensing, and diagnostics. IEEE Journal of Selected Topics in Quantum Electronics. 2016;**22**(3):1-14

[4] Schaefer S. Colorimetric Water Quality Sensing with Mobile Smart Phones. Vancouver, Canada: University of British Columbia; 2014

[5] Mutlu AY, Kılıç V. Machine learning based smartphone spectrometer for harmful dyes detection in water. In: 2018 26th Signal Processing and Communications Applications Conference (SIU); IEEE. 2018

[6] Özdemir GK, Bayram A, Kılıç V, Horzum N, Solmaz ME. Smartphone-based detection of dyes in water for environmental sustainability. Analytical Methods. 2017;**9**(4):579-585

[7] Intaravanne Y, Sumriddetchkajorn S. Android-based rice leaf color analyzer for estimating the needed amount of nitrogen fertilizer. Computers and Electronics in Agriculture. 2015;**116**:228-233

[8] Dutta S, Sarma D, Patel A, Nath P. Dye-assisted pH sensing using a smartphone. IEEE Photonics Technology Letters. 2015;**27**(22): 2363-2366

[9] Hossain MA, Canning J, Ast S, Cook K, Rutledge PJ, Jamalipour A. Combined "dual" absorption and fluorescence smartphone spectrometers. Optics Letters. 2015;**40**(8):1737-1740

[10] Wang Y et al. Smartphone spectrometer for colorimetric biosensing. Analyst. 2016;**141**(11):3233-3238

[11] Zhu H, Isikman SO, Mudanyali O, Greenbaum A, Ozcan A. Optical imaging techniques for point-of-care diagnostics. Lab on a Chip. 2013;**13**(1):51-67

[12] Zhu H et al. Cost-effective and rapid blood analysis on a cell-phone. Lab on a Chip. 2013;**13**(7):1282-1288

[13] Lillehoj PB, Huang M-C, Truong N, Ho C-M. Rapid electrochemical detection on a mobile phone. Lab on a Chip. 2013;**13**(15):2950-2955

[14] Coskun AF, Nagi R, Sadeghi K, Phillips S, Ozcan A. Albumin testing in urine using a smart-phone. Lab on a Chip. 2013;**13**(21):4231-4238

[15] Kim H, Awofeso O, Choi S, Jung Y, Bae E. Colorimetric analysis of saliva–alcohol test strips by smartphone-based instruments using machine-learning algorithms. Applied Optics. 2017;**56**(1):84-92

[16] McGonigle A et al. Smart-phone spectrometers. Sensors. 2018;**18**(1):223

[17] Salles M, Meloni G, De Araujo W, Paixão T. Explosive colorimetric discrimination using a smartphone, paper device and chemometrical approach. Analytical Methods. 2014;**6**(7):2047-2052

[18] Hossain MA, Canning J, Cook K, Jamalipour A. Optical fiber smartphone spectrometer. Optics Letters. 2016;**41**(10):2237-2240

[19] Chen J, Cai F, He R, He S. Experimental demonstration of remote and compact imaging spectrometer based on Mobile devices. Sensors (Basel, Switzerland). 2018;**18**(7)

[20] de Oliveira HJS et al. A handheld smartphone-controlled spectrophotometer based on hue to wavelength conversion for molecular absorption and emission measurements. Sensors and Actuators B: Chemical. 2017;**238**:1084-1091

[21] Yu H, Tan Y, Cunningham BT. Smartphone fluorescence spectroscopy. Analytical Chemistry. 2014;**86**(17):8805-8813

[22] Grasse EK, Torcasio MH, Smith AW. Teaching UV–vis spectroscopy with a 3D-printable smartphone spectrophotometer. Journal of Chemical Education. 2015;**93**(1):146-151

[23] Karisen H, Dong T. Illumination and device independence for colorimetric detection of urinary biomarkers with smartphone. In: 2016 IEEE 38th Annual International Conference of the Engineering in Medicine and Biology Society (EMBC); IEEE. 2016. pp. 5184-5187

[24] Jia M-Y, Wu Q-S, Li H, Zhang Y, Guan Y-F, Feng L. The calibration of cellphone camera-based colorimetric sensor array and its application in the determination of glucose in urine. Biosensors and Bioelectronics. 2015;**74**:1029-1037

[25] Jung Y, Kim J, Awofeso O, Kim H, Regnier F, Bae E. Smartphone-based colorimetric analysis for detection of saliva alcohol concentration. Applied Optics. 2015;**54**(31):9183-9189

[26] Zhang M, Leifer I, Hu C. Challenges in methane column retrievals from AVIRIS-NG imagery over spectrally cluttered surfaces: A sensitivity analysis. Remote Sensing. 2017;**9**(8):835

[27] Li-Chan E. The applications of Raman spectroscopy in food science. Trends in Food Science & Technology. 1996;**7**(11):361-370

[28] Solmaz ME, Mutlu AY, Alankus G, Kılıç V, Bayram A, Horzum N. Quantifying colorimetric tests using a smartphone app based on machine learning classifiers. Sensors and Actuators B: Chemical. 2018;**255**:1967-1973

[29] Bayram A, Horzum N, Metin AU, Kılıç V, Solmaz ME. Colorimetric bisphenol-A detection with a portable smartphone-based spectrometer. In IEEE Sensors Journal. 15 July, 2018;**18**(14):5948-5955

[30] Skandarajah A, Reber CD, Switz NA, Fletcher DA. Quantitative imaging with a mobile phone microscope. PLoS One. 2014;**9**(5):e96906

[31] Razdan S, Johannes J, Kuo RL, Bagley DH. The camera phone: A novel aid in urologic practice. Urology. 2006;**67**(4):665-669

[32] Bogoch II et al. Mobile phone microscopy for the diagnosis of soil-transmitted helminth infections: A proof-of-concept study. The American Journal of Tropical Medicine and Hygiene. 2013;**88**(4):626-629

[33] Tuijn CJ, Hoefman BJ, Van Beijma H, Oskam L, Chevrollier N. Data and image transfer using mobile phones to strengthen microscopy-based diagnostic services in low and middle income country laboratories. PLoS One. 2011;**6**(12):e28348

[34] Hong JI, Chang B-Y. Development of the smartphone-based colorimetry for multi-analyte sensing arrays. Lab on a Chip. 2014;**14**(10):1725-1732

[35] Gunda NSK, Naicker S, Shinde S, Kimbahune S, Shrivastava S, Mitra S. Mobile water kit (MWK): A smartphone compatible low-cost water monitoring system for rapid detection of total coliform and *E. coli*. Analytical Methods. 2014;**6**(16):6236-6246

[36] Berg B et al. Cellphone-based hand-held microplate reader for point-of-care testing of enzyme-linked immunosorbent assays. ACS Nano. 2015;**9**(8):7857-7866

[37] Long KD, Yu H, Cunningham BT. Smartphone instrument for portable enzyme-linked immunosorbent assays. Biomedical Optics Express. 2014;**5**(11):3792-3806

[38] Bremer K, Roth B. Fibre optic surface plasmon resonance sensor system designed for smartphones. Optics Express. 2015;**23**(13):17179-17184

[39] Morsy MK et al. Development and validation of a colorimetric sensor array for fish spoilage monitoring. Food Control. 2016;**60**:346-352

[40] Lopez-Ruiz N et al. Smartphone-based simultaneous pH and nitrite colorimetric determination for paper microfluidic devices. Analytical Chemistry. 2014;**86**(19):9554-9562

[41] Mutlu AY, Kılıç V, Özdemir GK, Bayram A, Horzum N, Solmaz ME. Smartphone-based colorimetric detection via machine learning. Analyst. 2017;**142**(13):2434-2441

[42] Sumriddetchkajorn S, Chaitavon K, Intaravanne Y. Mobile device-based self-referencing colorimeter for monitoring chlorine concentration in water. Sensors and Actuators B: Chemical. 2013;**182**:592-597

[43] Akkaynak D et al. Use of commercial off-the-shelf digital cameras for scientific data acquisition and scene-specific color calibration. Journal of the Optical Society of America A. 2014;**31**(2):312-321

[44] Pohanka M. Small camera as a handheld colorimetric tool in the analytical chemistry. Chemical Papers. September 2017;**71**(9):1553-1561

[45] Kılıç V, Alankus G, Horzum N, Mutlu AY, Bayram A, Solmaz ME. Single-image-referenced colorimetric water quality detection using a smartphone. ACS Omega. 2018;**3**(5):5531-5536

[46] Helfer GA, Magnus VS, Böck FC, Teichmann A, Ferrão MF, Costa AB d. PhotoMetrix: An application for univariate calibration and principal components analysis using colorimetry on mobile devices. Journal of the Brazilian Chemical Society. 2017;**28**(2):328-335

[47] Vashist SK, van Oordt T, Schneider EM, Zengerle R, von Stetten F, Luong JH. A smartphone-based colorimetric reader for bioanalytical applications using the screen-based bottom illumination provided by gadgets. Biosensors and Bioelectronics. 2015;**67**:248-255

[48] Coskun AF, Wong J, Khodadadi D, Nagi R, Tey A, Ozcan A. A personalized food allergen testing platform on a cellphone. Lab on a Chip. 2013;**13**(4):636-640

[49] Mudanyali O, Dimitrov S, Sikora U, Padmanabhan S, Navruz I, Ozcan A. Integrated rapid-diagnostic-test reader platform on a cellphone. Lab on a Chip. 2012;**12**(15):2678-2686

[50] Tseng D et al. Lensfree microscopy on a cellphone. Lab on a Chip. 2010;**10**(14):1787-1792

[51] Zhu H, Mavandadi S, Coskun AF, Yaglidere O, Ozcan A. Optofluidic fluorescent imaging cytometry on

a cell phone. Analytical Chemistry. 2011;**83**(17):6641-6647

[52] Yetisen AK, Martinez-Hurtado J, Garcia-Melendrez A, da Cruz Vasconcellos F, Lowe CR. A smartphone algorithm with inter-phone repeatability for the analysis of colorimetric tests. Sensors and Actuators B: Chemical. 2014;**196**:156-160

[53] Vora PL, Farrell JE, Tietz JD, Brainard DH. Digital Color Cameras–1–Response Models

[54] Zhang Y, Wu Y, Zhang Y, Ozcan A. Color calibration and fusion of lens-free and mobile-phone microscopy images for high-resolution and accurate color reproduction. Scientific Reports. 2016;**6**:27811

[55] Shankland S. How Android 5.0 lets you get raw for better photos. 2016. Available: https://www.cnet.com/news/android-lollipop-opens-high-end-photography-options/

[56] Iqbal Z, Eriksson M. Classification and quantitative optical analysis of liquid and solid samples using a mobile phone as illumination source and detector. Sensors and Actuators B: Chemical. 2013;**185**:354-362

Chapter 5

Fuel Quality Monitoring by Color Detection

Amanda Pereira Franco dos Santos, Kissya Kropf da Silva, Gisele Alves Borges and Luiz Antonio d'Avila

Abstract

The quality of automotive fuels—gasoline, diesel, biodiesel, and ethanol—is discussed in the light of their specifications in different countries and regions and in terms of their adulteration, which has such a prejudicial effect on the production chain, distribution chain, tax revenues, the environment, and end consumers. Different ways of adulterating automotive fuels are analyzed, as are the procedures for their detection. Several analytical methods for monitoring quality and detecting adulteration have been addressed in the literature, emphasizing the determination of properties such as density, distillation curve, octane rating, vapor pressure, etc., by means of physicochemical methods and chromatographic and spectrographic techniques, to the detriment of colorimetric methods. This chapter looks at colorimetric techniques designed for quality monitoring and the detection of adulteration in fuels, especially simple, quick, low-cost procedures with potential to be used in the field.

Keywords: quality, adulteration, gasoline, diesel, biodiesel, ethanol, methanol, glycerin, water, colorimetric tests, hydroxamic acid test, Schiff's reagent, solvatochromic effect, field test

1. Introduction

The quality of a fuel is associated with its fitness for use, the minimum requirements for which are given by its specifications, defined as a set of characteristics and their respective limits, which are required to ensure its good performance in engines. A comparison of the specifications for gasoline, ethanol, diesel, and biodiesel in different countries and regions is provided in the **Supplementary Material** to this chapter.

The failure of a fuel to meet any of its specifications is referred to as a nonconformity, which means the fuel is not fit for use. Nonconformity is not always due to adulteration, which is the deliberate and illegal addition of lower-cost substances to a fuel, usually with tax evasion, all with the purpose of increasing the profit margin by illicit means, resulting in reduced tax revenues, unfair competition, harm to the environment, and increased wear and tear of vehicle engines [1–5]. In fact, a fuel may be adulterated and yet stay within its specifications, in which case the only detrimental effect is on tax revenues [6]. It is a challenge to detect the adulteration of fuels because adulterants usually include compounds that are already present in the fuels themselves [2, 6].

The illegal practice of fuel adulteration may impair the fuel's quality while also resulting in increased environmental contamination by polluting gases and particulate matter because the combustion process is affected, leading to higher emissions of pollutants and compounds that cause acid rain, like NOx and SOx, as well as CO, which is highly asphyxiating [4, 7, 8].

The biggest detrimental effect of fuel adulteration is on the performance of the vehicle. In 2006, around 95% of vehicle repairs in São Paulo city, Brazil, were directly or indirectly attributable to poor fuel quality. Indeed, it has been estimated that fuel adulteration cost Brazil over 1 billion dollars that year, including 400 million dollars in lost tax revenues, affecting both local and federal budgets [9]. According to the Brazilian newspaper *Estadão*, a study done in 2017 by the Getúlio Vargas Foundation found that 4.8 billion reais is lost in the fuel sector every year to money laundering and tax evasion, with serious knock-on effects on the economy as a whole [10].

According to ANP, the agency responsible for monitoring fuel quality in the country, gasoline can be adulterated in multiple ways, such as by the addition of ethanol over the maximum permitted limit or the illicit addition of light and heavy aliphatic and aromatic solvents, which are themselves constituents of gasoline [6].

Gasoline and ethanol are both adulterated with methanol in Brazil, which is not allowed to exceed 0.5% v/v due to its toxicity, but whose low price makes it an attractive adulterant [11]. Meanwhile, the main nonconformity found in diesel in Brazil is a failure to comply with the correct quantity of biodiesel, which is set by ANP at 10% v/v [12]. Petroleum products like fuels are extremely important inputs for economies, industries, and basic everyday activities all around the world. Their importance to governments—whether for geopolitical reasons or because of armed conflicts or economic factors—means they tend to be taxed heavily, which raises their cost to consumers. As a result, adulteration has become a common practice in regions as diverse as South America and South Asia [13].

In India, kerosene is subsidized, which makes it particularly popular for adulterating gasoline, along with other products like naphtha, rubber solvents, aromatics, ethanol, light and heavy aliphatic solvents, and lubricants [2, 14, 15]. In 2018, the Petroleum Institute of East Africa announced that up to 75% of the 33 million liters of kerosene consumed every month ends up contaminating diesel and gasoline: only 5 million liters is actually used every month for lighting and cooking [16].

Cases of gasoline adulteration with acetone and sec-butyl acetate were reported in Vietnam and China in 2016, although this practice is less widespread in Vietnam now with the introduction of new limits for ketones in its gasoline specifications. In China, gasoline is often adulterated with substances like naphtha, ethanol, methanol, and even silicon oil or waste chemicals containing silicon [17].

In France, the most common form of adulteration consists of the addition of soybean oil or used cooking oil to diesel. In the last 5 years, the European Commission has invested heavily in preventing fuel adulteration [18].

According to Mani et al. (2017), the parameters that affect which adulterants are used in fuels—including considerations such as profitability, availability, and chemical compatibility—vary from place to place. The main adulteration methods are blending small quantities of middle distillate fuels like diesel and kerosene in gasoline; blending waste lubricants in gasoline and diesel, because the lubricants are expensive to dispose of in compliance with environmental standards; and adding small quantities of heavier fuel oils to diesel [3].

In Australia, the Department of the Environment and Energy regulates the Fuel Quality Standards Act 2000, which legislates and provides the standards on fuel quality in the country. In Europe, mandatory quality requirements for automotive gasoline and diesel are laid down by Directive 98/70/EC.

In the United States, ASTM International publishes standards on each type of fuel. Responsibility for setting the standards for fuel quality lies with the Committee D02 on Petroleum Products, Liquid Fuels, and Lubricants. Meanwhile, in Japan there are two standards for fuels: the mandatory standards set by the Central Environment Council, under the Ministry of the Environment, and the Japanese Industry Standards. The former are statutory requirements and regulate the Act on Quality Control for Gasoline and Other Fuels, which has been enforced since April 1996 [19]. **Tables 1–4** showing the gasoline, diesel, ethanol, and biodiesel specifications for each country or region are included in the **Supplementary Material**.

PROPERTY	AUSTRALIA	EUROPE			USA	JAPAN		SOUTH KOREA	BRAZIL
Spec Name/ Year	Fuel Quality Standards Act 2000 (2008)	Dir. 98/70/EC as amended (2009)	EN 228:2012 (2013)	EN 228:2012 (2013)	ASTM D 4814 (2014)	JIS K 2202 (2012)	JIS K 2202 (2012)	Petroleum and Alternative Fuels Business Act (2009)	Resolution ANP #40 (2013)
Source	Department of Environment	Dir. 98/70/EC as amended	EN 228:2012	EN 228:2012	ASTM International	Japanese Standards Association	Japanese Standards Association	Korea Petroleum Quality & Distribution Authority	ANP
Grade	ULP / PULP	Petrol	Unleaded Petrol	Unleaded Petrol E10	Unleaded	Regular / Premium	Regular (E) / Premium (E)	Regular / Premium	Regular (Type C) / Premium (Type C)
RON, min	91 / 95	95(1)	95(1)	95(1)		89 / 96	89 / 96	91 / 94	
MON, min	81 / 85	85	85	85					82 /
Antiknock index, calculated, min					(2)				87.0 / 91.0
Sulfur, ppm, max	150 / 50	10	10	10	80(3)	10	10	10	50
Lead, g/l, max	0.005	0.005	0.005	0.005	0.013(4)			0.013	
Manganese, g/l, max		2	2	2					
Benzene, vol%, max	1	1	1	1		1	1	0.7	1
Aromatics, vol%, max	42% pool average over 6 months with a cap of 45%	35	35	35				24(5)	35
Olefins, vol%, max	18	18	18	18				16(27)	25 (34)(35)
RVP à 37.8°C, kPa, min-max		60 max(6)(7)(8)	45-60 (class A) - 70-100 (class F1)(9)	45-60 (class A) - 70-100 (class F1)(9)	103 max(10)	44-65 (s) / 44-93 (w)	44 (s) / 55 (w)(11)	44-82(12)	69.0 max(13)
Density à 15°C, kg/m³, min-max			720-775	720-775		783 max	783 max		Note at 20°C
Distillation, T10, °C, max					70(14)	70	70	70	65
Distillation, T50, °C, min-max					77-121(14)(15)(16)	75-110	70-105 (s) / 65-105(w)	125 max	80
Distillation, T90, °C, min-max					190 max(14)	180 max	180 max	170 max	190
Distillation, E70, vol%, min-max			20-48 (class A) 22-50 (class F1)(9)	22-50 (class A) 24-52(class F1)(9)					
Distillation, E100, vol%, min-max		46 min	46-71	46-72					
Distillation, E150, vol%, min		75	75	75					
Distillation, FBP, °C, max	210		210	210	225	220	220	225	215
Distillation, Residue, vol%, max			2	2	2	2	2	2	2
Oxygen, wt%, min-max	2.7 max (no ethanol) 3.9 max (with ethanol)	3.7 max(17)	2.7 max	3.7 max		1.3 max	1.3-3.7	2.3 max(18)	
Methanol, vol%, max		3(19)	3	3				0.1 wt%	0.5
Ethanol, vol%, max	10	10(19)	5	10		3	10		(20)
Iso-propyl alcohol, vol%, max		12	(21)	12					
Iso-butyl alcohol, vol%, max		15	(21)	15					
Tert-butyl alcohol, vol%, max	0.5	15	(21)	15					
Ethers (5 or more C atoms), vol%, max	1 DIPE; 1 MTBE	22	(21)	22		7 MTBE	7 MTBE		
Phosphorus, g/l, max	0.0013		0(22)	0(22)	0.0013(4)			0.0013	
Induction period, minutes, min	360		360	360	240	240	240	480	360
Water and sediment, vol%, max								0.01	
Existent gum (solvent washed), mg/100ml, max	5		5	5	5	5	5	5	5
Copper corrosion, 3hr at 50°C, merit (class), max	1		1	1	1	1	1	1	1
Silver corrosion, merit (class), max					1				
Color						Orange	Orange	Yellow / Green	(23)
Appearance			Clear & bright	Clear & bright					LII(24)

(1) Member States may decide to continue to permit the marketing of gasoline with a minimum MON of 81 and a minimum RON of 91.
(2) Octane limits are set and regulated at the state level; the industry (R+M)/2 standard is generally 87/89/91+ for regular, midgrade and premium. Certification and posting of octane ratings regulated by Federal Trade Commission under 16 Code of Federal Regulations (CFR) 306.
(3) Per-gallon cap per EPA regulation (65 FR 6698; 2/10/00). The refinery average is 30 ppm.
(4) Leaded gasoline has been banned in the U.S. by EPA since 1996. Per EPA regulation (59 FR 7716; 2/16/94), no intentional addition of heavy metals allowed. While ASTM has no limit, EPA limits the phosphorus content of gasoline to a maximum of 0.0013 g/L. The regulations do not prohibit lead additives in aircraft, racing cars, and off-road farm & marine engines.
(5) Either aromatics 24 vol% max and olefins 16 vol% max, or aromatics 21 vol% max and olefins 19 vol% max.
(6) Vapor pressure determined for summer period.
(7) The summer period shall begin no later than May 1, and shall not end before Sept. 30. For Member States with low ambient summer temperatures the summer period shall begin no later than June 1 and shall not end before Aug 31.
(8) In the case of Member States with low ambient summer temperatures, where the derogation from 60 kPa, after the assessment and permission of the European Commission, is in effect, the maximum vapor pressure shall be 70 kPa. In the case of Member States with no low ambient summer temperatures, for gasoline containing bioethanol, after the assessment and permission of the European Commission, vapor pressure shall be 60 kPa plus the waiver specified in the Directive (0-8 kPa).
(9) Depends on volatility classes determined by the country's seasonal and geographical conditions. To relevant countries, volatility class A shall apply during summer starting no later than May 1 and ending not before Sept. 30. In countries with arctic conditions class B shall apply during summer, starting no later than June 1 and ending not before Aug. 31.
(10) ASTM advises to consult EPA for approved test methods for compliance with vapor pressure regulations. RVP varies by season and region. See EPA regulation (54 FR 11868; 3/22/89).
(11) If the ambient temperature is below -10°C, RVP min becomes 60 kPa.
(12) Hot climate (June - August): 60 kPa; Cold climate (October - March): 96 kPa.
(13) For the states of Rio Grande do Sul, Santa Catarina, Paraná, São Paulo, Rio de Janeiro, Espírito Santo, Minas Gerais, Mato Grosso, Mato Grosso do Sul, Goiás, and Tocantins, and for the Federal District, up to 7.0 kPa higher vapor pressure is permitted between April and November.
(14) Volatility requirements vary by season and region.
(15) Gasolines that may be blended with 1 to 10 vol% ethanol or all other gasolines whose disposition with ethanol blending is not known shall meet a minimum T50 of 77°C (170°F) prior to blending with ethanol. Gasolines that contain 1 to 10 vol% ethanol shall meet a minimum T50 of 66°C (150°F) after blending.
(16) Gasolines known from the origin to retail that will not be blended with ethanol may meet a minimum T50 of 66°C (150°F) for volatility classes D and E only. Gasolines meeting these limits are not suitable for blending with ethanol.
(17) Member States shall require suppliers to ensure the placing on the market of gasoline with a maximum oxygen content of 2.7% and a maximum ethanol content of 5% until 2013 and may require placing on the market of such gasoline for a longer period if they consider it necessary.
(18) Oxygen contained in MTBE, ETBE and Bio-ethanol.
(19) Stabilizing agents may be added.
(20) The level of anhydrous ethanol to be blended with regular gasoline to produce the gasoline-ethanol blend must comply with the existing regulations. At the time of writing, it stands at 27% vol.
(21) Volume blending restricted to 2.7 % (m/m) maximum oxygen content.
(22) In order to protect automotive catalyst systems, phosphorus containing compounds shall not be included in unleaded gasoline.
(23) From colorless to yellow if it does not contain a dye, although a dye can be added (up to 50 ppm), which can be any color but blue, which is reserved for aviation fuel.
(24) Clear and free of suspended matter.

Table 1.
Gasoline specification.

PROPERTY	AUSTRALIA	EUROPE		USA	JAPAN	SOUTH KOREA	BRAZIL
Spec Name / Year	Fuel Standard (Automotive Diesel)	Dir. 98/70/EC as amended (1) (2009)	EN 590 (2013)	ASTM D 975 (2014)	JIS K 2204 (2007)	Petroleum and Alternative Fuels Business Act	Resolution ANP #50 (2013)
Source	Department of the Environment	Dir. 98/70/EC as amended	EN 590:2013	ASTM International	Japanese Standards Association	Korea Petroleum Quality & Distribution Authority	ANP
Grade		Diesel	Diesel	N° 1 - D S15 / N° 2 - D S15	Class Special 1 / Class 1 / Class 2 / Class 3 / Class Special 3	Automotive Diesel	S10 / S500
Cetane number, min	51.0	51	51 (temperate) / 47-49 (arctic & severe winter)[1]	40	50 / 50 / 45 / 45 / 45[2]	52[3]	48 / 42[4]
Cetane index, min	46.0		46 (temperate) / 43-46 (arctic & severe winter)[1]	40(11)	50 / 50 / 45 / 45 / 45[2]	52[3]	
Sulfur, ppm, max	10	10	10	15[5]	10	10	10 / 500
Polyaromatics, wt%, max	11	8	8			2-5	11 /
Density à 15°C, kg/m³, min-max	820 - 850		820-845 (temperate) / 800-840 (arctic & severe winter)[1]		860	815-835	815.0-850.0 à 20°C (22) / 815.0-865.0 à 20°C
Viscosity at 40°C, cSt, min-max	2.0 - 4.5		2.0-4.5 (temperate) / 1.2-4.0(arctic & severe winter)[1]		2.7 / 2.7 / 2.5 / 2 / 1.7	1.9-5.5	2.0-4.5 (mm²/s) / 2-5 (mm²/s)
Distillation, T10, °C, min - max							180 /
Distillation, T50, °C, min - max							245-295 / 245-310
Distillation , T85, °C, min - max							/ 360
Distillation, T90, °C, min - max				282-338[6]	360 / 360 / 350 / 330[7] / 330	360	
Distillation, T95, °C, max	360	360	360				370 /
Distillation, E180, vol%, max			10				
Distillation, E250, vol%, max			<65				
Distillation, E340, vol%, min			95[8]				
Distillation, E350, vol%, min			85				
Flash Point, °C, min	61.5		55	52[6]	50 / 50 / 50 / 45 / 45	40	38
Carbon residue 10%, wt%, max	0.2		0.3	0.35	0.1	0,15	0.25
Cold Filter Plugging Point, °C, max			+5 (class A temperate) to -44(arctic & severe winter)[1]		/ -1 / -5 / -12 / -19	-18	[1]
Cloud Point (CP), °C, max			-10 to -34[9]				
Pour Point (PP), °C, max					5, -2.5, -7.5, -20, -30	-23.0 (w) / 0.0 (s)	
Water and sediment, vol%, max	0.05			0.05		0.02	0.05
Water, mg.kg⁻¹, max			200				200 / 500
Ash, wt%, max	0.01		0.01	0.01		0.02	0.010
Lubricity, HFRR wear scar diam à 60°C, micron, max	0.460		460	520		400	
Copper corrosion, 3hr at 50°C, merit (class), max	1		1	3		1	1
Oxidation stability, mg/100ml, max	25		5				2.5 /
FAME content, vol%, max	5.0	7[10]	7	5		5	10[11]
Conductivity at ambient temp, pS/m, min	50			25			25

(1) Depends on climate rating.
(2) Cetane number or cetane index may be used.
(3) Winter (Nov. 15 - Feb. 18): 48.
(4) Alternatively, the determination of the cetane number using the method described in Brazilian standard NBR 14759 (ASTM D4737) is permitted when the product does not contain a cetane improver, with a minimum limit of 45. In cases of nonconformity, cetane testing should be done. The producer and importer must include information on any use of a cetane improver in the quality certificate. Note that the cetane number does not translate into the quality of ignition of diesel containing biodiesel and/or cetane booster.
(5) Other limits may apply to selected areas.
(6) When a cloud point is less than -12°C is specified, it is permitted and normal blending practice to combine Grades No.1-D and No.2-D to meet the low temperature requirements. In that case, the minimum flash point shall be 38°C, the minimum viscosity at 40°C shall be 1.77 cSt, and the minimum 90% recovered temperature shall be waived.
(7) T90 should be 350°C max if viscosity is 4.7 cSt max.
(8) Only applicable to countries with arctic or severe winter conditions.
(9) Only applicable to countries with arctic or severe winter conditions.
(10) Either the specification for minimum cetane index or that for maximum total aromatics must be met.
(11) Only applicable for a diesel-biodiesel blend.

Table 2.
Diesel fuel specification.

PROPERTY	AUSTRALIA	EUROPE	USA	BRAZIL
Spec Name / Year	Fuel Standard (Ethanol E85) Determination (2012)	CEN/TS 15293 (2011)	ASTM D 5798 (2013)	Resolution ANP #19 (2015)
Source	Department of the Environment	CEN/TS 15293	ASTM International	ANP
Grade	E85	E85	E85-Class 1 E85-Class 2 E85-Class 3 E85-Class 4	EAC (Anhydrous Ethanol)/ EHC (Hydrous Ethanol)
RON, min	100	104[1]		
MON, min	87	88[1]		
Sulfur, ppm, max	70	10	80	Note
Lead, g/l, max	0.005		[2]	
Benzene, vol%, max	0.35			
RVP à 37.8°C (100°F), kPa, min-max	38-65	35-60 (s) / 50-80 (w)(2)	38-62 / 48-65 59-83 / 66-103	
Density at 15°C , kg/m³, min-max		760-800		791.5 máx (11) à 20°C / 805.2-811.0(11) à 20°C
Distillation, FBP, °C, max	210			
Methanol, vol%, max	0.5	1.0	0.5	0.5
Ethanol, vol%, min-max	70-85	70-85 (s) / 50-75 (w)(2)	51-83	98 / 94.5
Ethers (5 or more C atoms), vol%, max	1	11		
C3-C5 alcohols, ppm, max	2 vol%[4]	6	2 vol%[4]	
Phosphorus, g/l, max	0.0013	0.00015	[5]	
Oxidation stability, minutes, min	360	360		
Water, vol%, max	1 wt%	0.4	1 wt%	0.7 (wt%) / 7.5 (wt%)
Existent gum (solvent washed), mg/100ml, max	5	5	5	
Existent gum (solvent unwashed), mg/100ml, max			20	
Chloride, inorganic, ppm, max	1	1.2		1
Copper, ppm, max	0.1	0.1	0.07	0.07
Copper corrosion, 3h, 50°C, merit (class), max		1		
Color		Clear & bright		[6] / [7]
pH, min-max	6.5-9.0	6.5-9.0	6.5-9.0	/ 6.0 a 8.0
Acidity, wt%, max	0.006	0.005[8]	0.005[8]	
Electrical Conductivity, µS/m, max		1.5		300
Sulfate, ppm, max	4	4		4
Appearance		Clear and bright	Clear and bright	LII[9]
Total acidity, Max, mg/L				30
Alcohol content, %wt				99.3 min / 92.5-94.6
Residue by evaporation, mg/100 mL, max.				5
Hydrocarbon content, %vol, max				3
Iron content, max. (mg/Kg)				5
Sodium content, max. (mg/Kg)				2

* *Until 2004, the main use of ethanol in Japan was for beverage production (Buainain AM. Cadeia Produtiva de Agroenergia. 2007. p. 32).*
** *At the time of writing, South Korea uses around 140 million gallons of ethanol a year, but just for beverages and industrial uses. (https://grains.org/south-korean-trade-team-focuses-on-fuel-ethanol-potential/).*
(1) Recommended value.
(2) Lead is not permitted to be added, according to Federal Regulations; the lead limit for gasoline is 0.013 g/L.
(3) Summer: Class A May 1 - Sept. 30, Winter: CEN notes that each country must choose which climate classes to use for other periods of the year.
(4) C3-C8.
(5) Phosphorus may not be added, according to federal regulations; the phosphorus limit in gasoline is 0.0013 g/L.
(6) Orange after the addition of the dye specified in Table VII of these technical regulations.
(7) May not be orange or blue, which are for anhydrous ethanol and aviation fuel, respectively.
(8) As acetic acid.
(9) Clear and free of suspended matter

Table 3.
Ethanol fuel specification.

PROPERTY	ÁUSTRIA	CZECH REPUBLIC	FRANCE	GERMANY	ITÁLY	USA	EUROPE	COLÔMBIA	BRAZIL
Spec Name	ON	CSN	Journal Oficial	DIN	UNI	ASTM	EN 14214	ICONTEC	
Appearance									LII(1)
Total Contamination, max. mg Kg^{-1}		24		20			24	24	24
Free Glycerin, max. wt%	0.02	0.02	0.02	0.02	0.02	0.02	0.02	0.02	0.02
Monoglycerides, max. wt%			0.8	0.8	0.8		0.8	0.8	0.8
Diglycerides, max. wt%			0.2	0.4	0.2		0.2	0.2	0.2
Triglycerides, max. wt%			0.2	0.4	0.1		0.2	0.2	0.2
Total Glycerin, max. wt%	0.24	0.24	0.25	0.25		0.24	0.25	0.25	0.25
Carbon Residue, max. wt%	0.05	0.05		0.05		0.05	0.3(2)		0.05(2)
Fatty Acid Methyl Esters, min. wt%			96.5		98		96.5	96.5	96.5
Flash Point, min. °C		100	110	100	110	130	120	120	100
Methanol or ethanol, max. wt%	0.2		0.1	0.3	0.2		0.2	0.2	0.2
Sulfated Ash, max. wt%	0.02	0.02		0.02			0.02	0.02	0.02
Sodium + Potassion, max. mg.Kg^{-1}							5		5
Density at 15°C, min-max. Kg m^{-3}	850-890	870-890	870-900	870-900	860-900		860-900	860-900	850-900(3)
Viscosity at 40°C, min-max. mm^2 s^{-1}	3.5-5.0	3.5-5.0	3.5-5.0	3.5-5.0	3.5-5.0	1.9-6.0	3.5-5.0	1.9-6.0	3.0-6.0
Iodine, max. g I2 100g^{-1}	120		115	115			120		Note
Distillation T90, max. °C			360		360	360		360	
Cetane Number, min-max.	49	48	49	49		47	51	47	Note
Clould Point, °C						Anotar			
Cold Filter Plugging Point, min-max. °C	0/-15	-5		0/-20			Por região		By region
Poor Point, min-max. °C			-10		0/-5		Por região	3	

(1) Clear and free of suspended matter
(2) Determination of carbon residue on 10 % (volume) distillation residues
(3) Specification 20°C;

Table 4.
Biodiesel fuel specification.

Analytical methods for monitoring gasoline and diesel adulteration were the target of a recent review [2], emphasizing the determination of their different properties, such as density, distillation curve, octane rating, vapor pressure, etc., using physicochemical methods and chromatographic and spectroscopic techniques. Nondestructive analytical methods for identifying off-spec levels of biodiesel in diesel-biodiesel blends and adulteration of the blend have also been covered in a recent review [20], focusing on first-order multivariate calibration models. In this chapter, different colorimetric techniques that can be used in fuel quality control are addressed, especially simple, rapid, low-cost methods that have the potential capacity to detect fuel adulteration in the field.

2. Colorimetric methods for monitoring fuel quality in laboratories

The colorimetric assays used on a routine basis in fuel quality control laboratories in Brazil are visual color, ASTM color, and the copper strip corrosion test.

In Brazil, hydrous ethanol should be colorless, while anhydrous ethanol has an orange dye added to it. This difference is designed to prevent the kind of fraud known as "wet ethanol," whereby fraudsters acquire anhydrous ethanol directly from the plants that produce it, evading taxes, then mix it with water, and sell it on as hydrous ethanol. Regular gasoline, additized gasoline, and premium gasoline can be told apart visually by their color: regular gasoline ranges from colorless to yellow, while the other two are colored with a dye, which may be of any color but blue (which is reserved for aviation fuel), but which is normally green for additized and purple for premium [11, 21]. All the dye does is enable the two products to be distinguished from one another; it has no effect on their specifications. Meanwhile, diesel S10 (containing 10 mg kg^{-1} sulfur) is naturally yellow in color, while S500 (500 mg kg^{-1} sulfur) is dyed red to prevent incorrect refueling with this diesel, whose high sulfur level means it could cause damage to vehicles with more sophisticated engine technologies like selective catalytic reduction or exhaust gas recirculation systems.

The standard test method for the ASTM color of petroleum products (ASTM 1500:2012) is used on samples of diesel S10 (which contains no dye). The test involves transferring a sample of the product to a test tube and comparing its color with the color of standard colored glass disks, which represent a range of values from 0.5 to 8.0. A standard light source is used in this test [22]. Corgozinho et al. have developed a method using partial least squared regression with molecular absorption spectrophotometry to determine the ASTM color of diesel in the 0–6.5 range, offering greater precision than the standard colorimetry test [23].

Corrosiveness is determined using the copper strip test in Brazil (ABNT NBR 14359), the USA (ASTM D130), and Europe (EN ISO 2160). It consists of immersing a standard copper strip in a sample of biodiesel in a corrosion bath for 3 hours at 50°C. The strip is then rinsed in solvent and compared with an ASTM copper strip corrosion standard, which has different colors that correspond to different levels of corrosion, classified from 1 to 4 [24–26].

3. Colorimetric methods for monitoring fuel quality with potential for use in the field

3.1 Biodiesel

3.1.1 In a diesel-biodiesel blend

The level of biodiesel in a diesel-biodiesel blend is determined in a laboratory by observing the mid-infrared absorption of the C=O bond present in the fatty acid methyl esters (FAMEs) that constitute the biodiesel. Although this test gives satisfactory results, it cannot be used in the field, which is a major drawback, as fuel distributors need to test the fuel when they are adding biodiesel to diesel on a commercial scale. Furthermore, the accuracy of the results will be compromised if the biodiesel itself is adulterated with vegetable oil or animal fat, whose constituent triglycerides also have C=O bonds.

The conversion of esters into hydroxamates by their reaction with hydroxylamine [27–29], forming complexes of a red to violet color caused by the reaction with $Fe3^{+}$, has been used to confirm esters in organic analysis procedures. **Figure 1** shows the reaction by which this complex is formed.

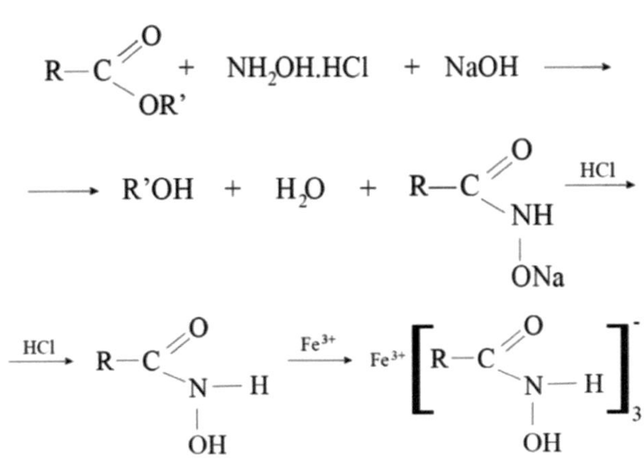

Figure 1.
Chemical reactions involved in the test for esters known as the hydroxamic acid test, used to test for the presence of biodiesel in fossil diesel. Adapted from Silva et al. [29].

Silva et al. used this test to develop a new spectrophotometric method to determine biodiesel levels in diesel-biodiesel blends ranging from B0 (pure fossil diesel) to B5 (fossil diesel containing 5% biodiesel), in which the esters from the biodiesel present in the mixtures react with hydroxylamine chlorhydrate in an alkaline solution, forming alkaline salts from the hydroxamic acid, followed by acidification, to form hydroxamic acid, followed by a reaction with Fe^{3+} ions. The complex formed, the intensity of whose color is proportional to the biodiesel content, is extracted with *n*-hexane/*n*-heptane, forming an upper phase with a yellow, orange, or red color, depending on the biodiesel concentration, as shown in **Figure 2**. Absorbance can be measured using spectrophotometry at 420–440 nm. The linearity, limit of detection, limit of quantification, accuracy, selectivity, and specificity results indicate that the proposed method is adequate for analyses of biodiesel in diesel-biodiesel blends. A simplified method was proposed for qualitative analyses for use at filling stations [29, 31].

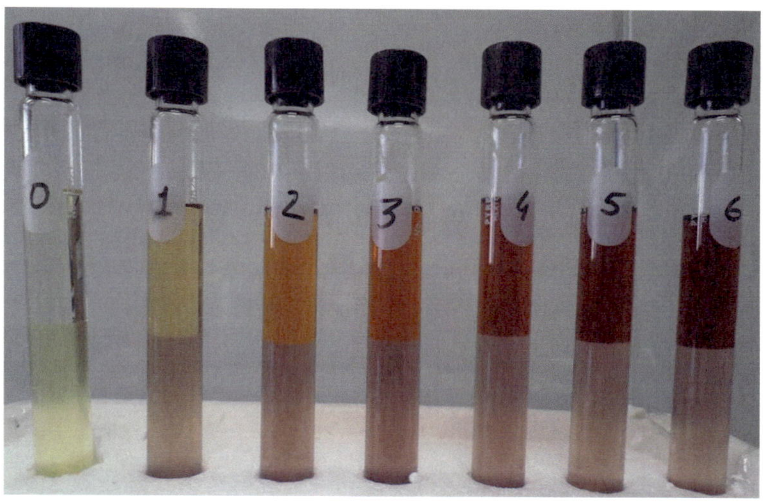

Figure 2.
Color scale of samples of fossil diesel containing biodiesel at 0–6% v/v, showing intensities proportional to the concentrations of biodiesel in the diesel, for semiquantitative analysis observable by the naked eye. Source: Santos [30].

Using the same reaction, Santos developed a simple, practical, rapid method that could be used in the field for semiquantitative identification by the naked eye of biodiesel in diesel-biodiesel blends ranging from 0% to 6% v/v biodiesel [30]. A similar method was developed by Leite and Fernandes for the 0–7% range [32]. What sets these methods apart from their predecessors is the separation of the biodiesel from the blend before the hydroxamic acid test. To do this, solid-phase extraction is conducted using a chromatographic column adapted from a 3mL disposable plastic syringe into which around 0.4 g silica gel is inserted, supported by a cotton plug (see **Figure 3a**). The diesel-biodiesel blend is introduced as if in frontal chromatography, pushed by a plunger, generating a less polar diesel fraction (F1) followed by a more polar fraction containing biodiesel (F2), displaced by the addition of ethanol (see **Figure 3b**).

By using seven reference samples of the diesel-biodiesel blend containing 0, 1, 2, 3, 4, 5, and 6% (v/v) biodiesel, the concentrations of which were confirmed by the laboratory reference method [33], a standard table of colors was prepared that could be used for the semiquantitative analysis of the percentage of biodiesel, as in **Figure 2**, since the intensity of the color formed by the ferric hydroxamate complex is proportional to the level of biodiesel present in the sample. By comparing the observed colors with the standard color table, it was possible to determine the

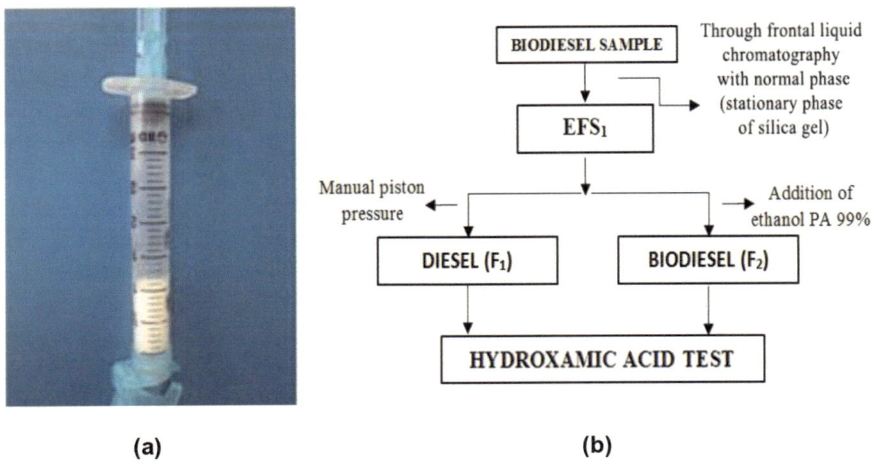

Figure 3.
(a) Chromatographic column adapted from a disposable plastic syringe; (b) solid-phase extraction of the biodiesel from the diesel-biodiesel blend before undergoing the hydroxamic acid test.

biodiesel concentration semiquantitatively with the naked eye, with a relative error of around 1%. This could be done in the field to find out whether samples are off spec. The proposed method was used on 33 samples of diesel containing different biodiesel concentrations and was found to be equivalent to the laboratory reference method (EN14078) by Student's t-test, with 95% confidence [30].

Another method—also adaptable for use in the field—has been developed for determining biodiesel in a diesel-biodiesel blend and for identifying the presence of vegetable oil in this blend. After solid-phase extraction using a silica stationary phase (EFS_1), the fraction composed of biodiesel and potentially vegetable oil is then put through another solid-phase extraction with an aminopropyl stationary phase (EFS_2) to separate the biodiesel from any vegetable oil (**Figure 4a**). For use in the field, the stages involving a manifold for solid-phase extraction and nitrogen-supported solvent evaporation are replaced by the manual use of plungers in the solid-phase extraction cartridges, which has yielded viable preliminary results. **Figure 4b** illustrates the procedures that can be done in the field to obtain the complexes. **Figure 5** also illustrates the use of the hydroxamic acid test on both fractions to confirm the presence of esters, which has also proven satisfactory for indicating adulteration with 1% or more vegetable oil in diesel containing 5% biodiesel [32].

Any sample identified in the field as potentially adulterated with vegetable oil could be sent to a laboratory for confirmation via more precise analytical techniques, like high-performance liquid chromatography. As such, both methods (semiquantitative identification of the percentage of biodiesel in diesel and the identification of any vegetable oil in the blend) are potentially effective for field analyses not only for their ease of use but also for their speed. Furthermore, the fact that the methods use low-cost, low-toxicity materials and reagents makes them even more attractive.

The sequence of operations for the hydroxamic acid test, illustrated in **Figure 5**, is easily adapted for field analysis.

Costa has also developed an alternative method for determining the level of biodiesel in fossil diesel using the hydroxamic acid test in association with image processing and a colorimeter and compared it with the standard method (EN 14078). After the hydroxamic acid test, the images were photographed, and the

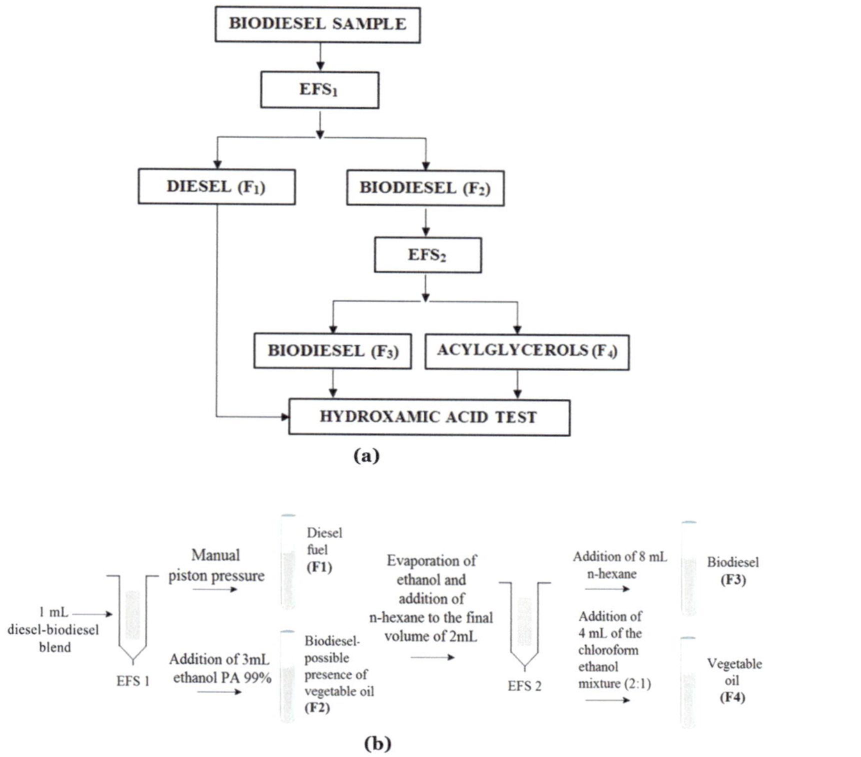

Figure 4.
(a) Sequence of operations for solid-phase extraction to separate biodiesel from diesel-biodiesel blends in a silica column (EFS_1) and to separate any vegetable oil or animal fat contaminating the biodiesel in an aminopropyl column (EFS_2). (b) Field test made up of solid-phase extraction followed by hydroxamic acid test.

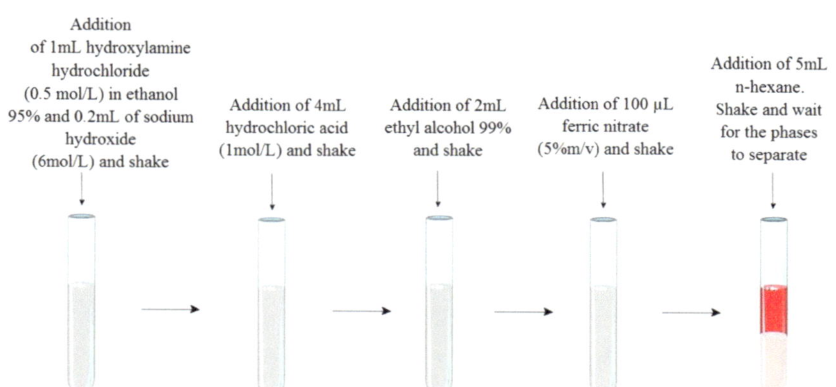

Figure 5.
Schematic representation of hydroxamic acid colorimetric test. Adapted from Costa [34].

Colorsys program was used to create calibration curves based on color vs. biodiesel concentration, enabling the determination of biodiesel content by the intensity of the color. It was found that the hydroxamic acid test associated with image processing was adequate for detecting and quantifying the biodiesel content of diesel-biodiesel blends because the results given by the method were statistically similar to those given by the reference method using infrared spectroscopy (EN 14078). Furthermore, it is easily adapted for field analysis (**Figure 6**) [34].

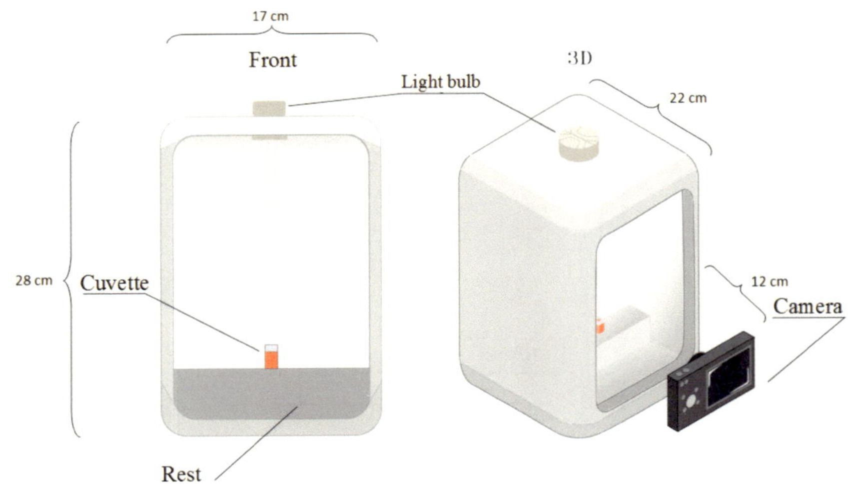

Figure 6.
Schematic representation of procedure for taking pictures of samples (front and perspective views). Source: Costa [34].

Solvatochromism is the term used to describe the phenomenon of color change of a solute when it is dissolved in different solvents, resulting in a change in the position of absorption or emission of its spectroscopic band. When these changes of position occur in the visible range, they exhibit different colors that can be detected by the naked eye and can be used as sensors to determine the composition of mixtures [59–61].

Fong and Xue have developed an optical sensor to detect biodiesel in diesel-biodiesel blends based on the solvatochromic behavior of the dye called Nile Blue chloride, which turns blue when dissolved in ethanol or methanol and pink when in contact with biodiesel. **Figure** 7 illustrates the change in color, enabling the direct, rapid detection of variations from 0.5 ppm to 20% v/v biodiesel in diesel-biodiesel blends [35].

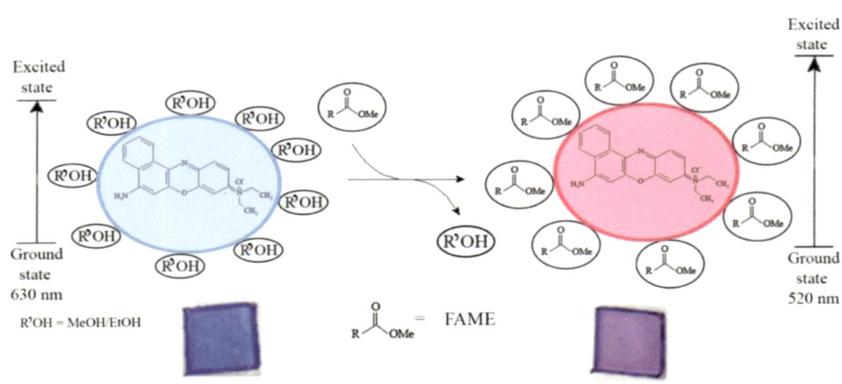

Figure 7.
Change in color of Nile Blue chloride from blue (without diesel) to pink (exposed to biodiesel), due to solvatochromic effects. Adapted from Fong and Xue [35].

As Nile Blue chloride is not soluble in nonpolar solvents, like diesel, it was encapsulated in an ethyl cellulose polymer, dissolved in methanol or ethanol in the form of a fine film that turns blue in alcohols, with absorbance at 610 nm, and pink when in contact with biodiesel, with absorbance at around 520 nm. Exposing the sensor containing the Nile Blue chloride to diesel resulted in no change in color or change in its stability. When it was exposed to biodiesel, it changed from blue to pink, which can be attributed to the alcohols surrounding the sensor being replaced by FAMEs from the biodiesel, resulting in a solvatochromic effect. The sensor provides a quick, direct method compared with methods involving gas chromatography or infrared spectrophotometry, presenting a linear response between color intensity and biodiesel concentration for low concentrations, between 0.5 and 30 mg kg^{-1}. For high concentrations of biodiesel (up to 20% v/v), absorbance may be linearly adjusted by a logarithmic function [35].

3.1.2 Free and total glycerin

Free or total glycerin in biodiesel can be determined by spectrophotometry, since the reactions shown in **Figure 8** result in a colored compound, as initially described by Greenhill [37]. Several publications have used similar methods, taking advantage of kits used to analyze triglycerides in blood [36, 38–40]. In the presence of adenosine-5'-triphosphate (ATP) and the enzyme glycerol kinase (GK), glycerol produces diphosphate adenosine-5'diphosphate (ADP) and glycerol-3-phosphate (G-3-P; reaction 1), which, in the presence of glycerol phosphate oxidase (GPO) and oxygen, produces dihydroxyacetone phosphate (DAP) and hydrogen peroxide (H_2O_2; reaction 2). Then, in the presence of a peroxidase, an oxygen acceptor (OA), and 4-aminoantipyrine (4-AAP), water and a colored compound (CC) are formed (reaction 3):

Glycerol + → G-3-P + ADP (in the presence of GK) **(Reaction 1)**
G-3-P + O_2 → ATP DAP + H_2O_2 (in the presence of GPO) **(Reaction 2)**
H_2O_2 + 4-AAP + OA → CC + H_2O (in the presence of peroxidase) **(Reaction 3)**

Figure 8.
Glycerol reactions generating a colored compound analyzable by spectrophotometry. Adapted from Muniz et al. [36].

To determine total glycerin, it is necessary to use solid-phase extraction in an aminopropyl cartridge to separate the biodiesel (FAMEs) from the fraction containing the free glycerin (FG), monoacylglycerols (MAGs), diacylglycerols (DAGs), triacylglycerols (TAGs), and combined glycerin, as shown in **Figure 9a** [36]. To determine free glycerin, the biodiesel (FAME) is separated and isolated from the fraction containing the free glycerin (FG) by solid-phase extraction in a silica cartridge (see **Figure 9b**).

3.2 Ethanol

Budag et al. have used four dyes as solvatochromic probes to study the effect of adding ethanol to pure gasoline. The dye 2,6-di(4-tert-butylphenyl)-4-[2,4,6-tri (4-tert-butylphenyl)pyridinium-1-yl]phenolate (t-Bu_5RB), for example, has a greenish blue color in gasoline, is violet in ethanol, and is bluish green in a mixture of 25% ethanol in gasoline, which means ethanol in gasoline can be detected by the naked eye. The development of analytical methods to determine fuel quality based on these promising probes has been discussed. Based on the spectroscopic determination of the maxima on the UV-visible spectra, the respective transition energy can be calculated, which is related to the composition of the fuel, opening up the potential for the development of an analytical method [41].

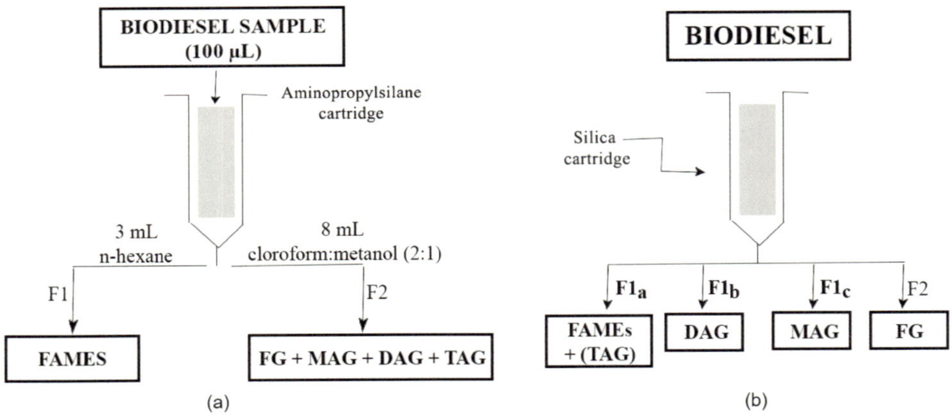

Figure 9.
(a) Solid-phase extraction in aminopropyl cartridge to separate and isolate total glycerin, made up of fractions of free glycerin (FG) and combined glycerin, monoacylglycerols (MAGs), diacylglycerols (DAGs), and triacylglycerols (TAGs). Source: Muniz et al. [36]. (b) Solid-phase extraction in silica cartridge to separate and isolate FG from the fractions containing FAME + TAGs, DAGs, and MAGs. For example, for 0.150 mL biodiesel, F_{1a} is eluted with 15 mL petroleum ether, (F_{1b}) with 40 mL of a mixture of 35% ethyl ether and 65% petroleum ether, (F_{1c}) with 40 mL ethyl ether, and (F_2) with 12 mL ethanol. Source: Serralvo Neto et al., 2018 (FR 1872032) [40].

El Seoud et al. have used the same dye in analytical chemical experiments for undergraduates, in the analysis of diesel-ethanol blends, also observing a solvatochromic effect, with colors varying as the composition of the mixture changes, as shown in **Figure 10** [42].

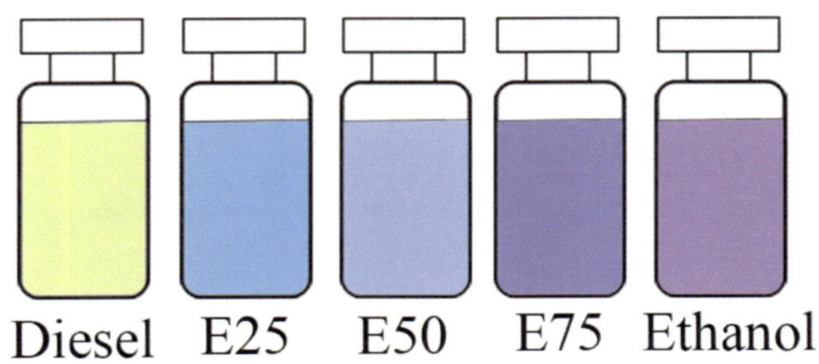

Figure 10.
Color variation of solutions of t-Bu_5RB and pure diesel and ethanol and in solutions of ethanol (%v/v) in diesel. Adapted from El Seoud et al. [42].

Galgano et al. have also used the solvatochromic dye 2,6-bis[4-(tert-butyl)phenyl]-4-{2,4,6-tris[4-(tert-butyl)phenyl]-pyridinium-1-yl}phenolate (t-Bu)5RB) to make quantitative analyses of bioethanol and mixtures of bioethanol with gasoline and water in an experiment based on a constructivist approach for undergraduates in a chemistry class [43].

Solutions of the dye in ethanol-water and anhydrous ethanol-gasoline mixtures evidenced solvatochromic effects, with the color changing according to the composition of the mixture, as shown in **Figure 11.**

The results of using solvatochromic effects of different dyes in mixtures, especially involving additized or altered fuels, seem extremely promising for the development of alternative colorimetric methods with field applications.

3.2.1 Methanol

The maximum methanol content permitted in ethanol and gasoline in Brazil is 0.5% v/v, and the reference method for its detection uses gas chromatography [44].

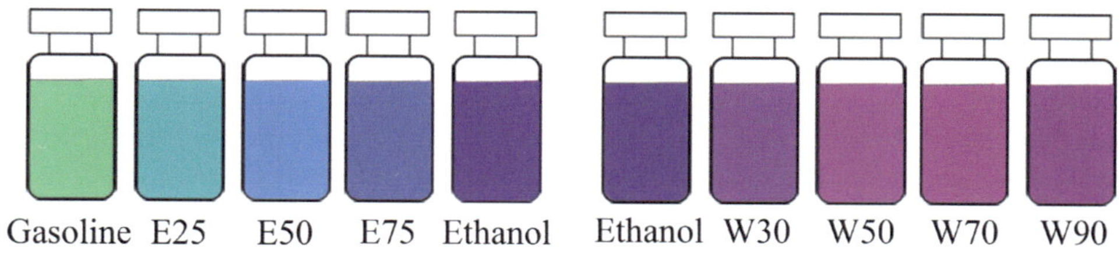

Figure 11.
Range of colors of solutions of the dye in binary mixtures of gasoline-ethanol (pure gasoline, E25, E50, E75, pure ethanol) and binary mixtures of ethanol-water (pure ethanol, W30, W50, W70, W90). Adapted from Galgano et al. [43].

In view of cases of ethanol adulteration with methanol identified in the country, a colorimetric kit has been designed for use in the field to identify this form of adulteration of anhydrous ethanol and hydrous ethanol and in gasoline with ethanol [45–47].

The procedure is based on visual detection of a violet blue color as a result of the reaction of aldehydes with Schiff's reagent used in qualitative organic analysis to identify functional aldehyde groups [48, 49]. The color is obtained using the aforementioned colorimetric kit by the following reactions:

1. Selective oxidation of the methanol adulterant to formaldehyde in the presence of potassium permanganate in an acid medium

2. Addition of oxalic acid in an acidic medium to reduce the excess of potassium permanganate, leaving the solution colorless

3. Formation of Schiff's reagent by the reaction of *p*-rosaniline with sodium metabisulfite in an acid medium

4. Addition of Schiff's reagent and observation of a violet blue color

The nonappearance of a color indicates the absence of methanol. When the solution turns violet blue, this is a qualitative indication of the presence of methanol in the fuel. A comparison of the intensity of the color with that of a reference solution containing 0.5% v/v methanol (the maximum permitted) enables conformity/non-conformity with specifications to be determined in the field.

Figure 12 shows the results of the samples of hydrous ethanol containing 0, 0.3, 0.5, and 1.0% methanol, showing the photos of the solutions after the test sequence. The higher the concentration of methanol in the sample, the stronger the blue of the solution. As expected, the blank, containing just ethanol, is colorless.

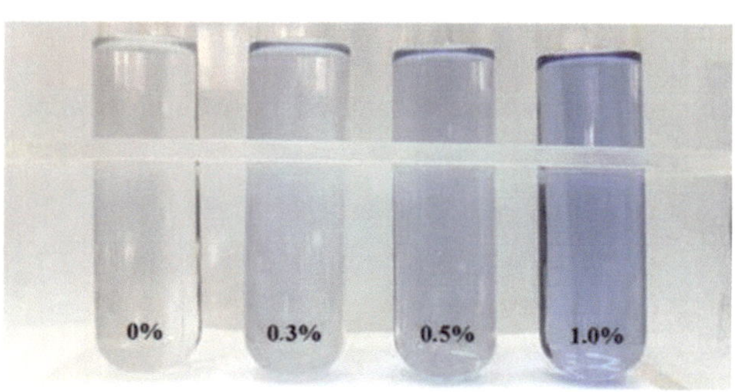

Figure 12.
Schiff test applied to samples of pure ethanol and ethanol contaminated with 0.3, 0.5, and 1.0% methanol.

3.2.2 Water

In Brazil, ANP Resolution #19, of 2015, states that hydrous ethanol fuel may contain up to 7.5% water [50]. Giordano et al. have developed a simple, rapid, portable, low-cost method that can be used in the field to indirectly determine water in ethanol fuel. The system is based on a colorimetric reaction between ethanol and ceric oxide (Ce(IV)), as shown in **Figure 13.** When a solution of ammonium cerium (IV) nitrate ($[(NH_4)_2Ce(NO_3)_6]$, which is yellow in color, is mixed with ethanol, an intense orangey-red color immediately appears because of the formation of Ce(IV)-ethanol (1:1) complexes [51].

$$Ce(IV)_{(yellow)} + C_2H_5OH_{(colourless)} \xrightarrow{K_{eq}} Ce(IV) \cdot C_2H_5OH_{(orange\text{-}red)}$$

Figure 13.
Colorimetric reaction between the reagent Ce(IV) and ethanol. Adapted from Giordano et al. [51].

The device is made up of mechanically connected layers of acrylic, a sample reservoir, a light source, and a detector. A solution containing the reagents is added to the reservoir, together with aqueous solutions of ethanol (standards or real samples) prepared at 0.2–5.0% v/v and transferred to the reservoir. The volume introduced is approximately 70 μL. As soon as the solutions are mixed, the reactions take place, as indicated by a change in color from yellow to orangey red.

Real samples of ethanol obtained from filling stations in Brazil were tested. The water concentrations were compared with those obtained by Karl Fischer titration (ASTM E203, reference method), and the data obtained by the two techniques were found to be statistically equivalent by Student's t-test, at a 95% confidence level [51, 52].

3.3 Gasoline

3.3.1 Methanol

The method for determining methanol in gasoline containing ethanol, applicable in the field, was described in 3.2.1.

3.3.2 Detergent-dispersant additives

Gasoline is additized with detergent-dispersant additives, which are designed to reduce the formation of deposits in engines and valves, enhance car engine efficiency, and ensure a cleaner combustion process, all of which has a positive impact on the environment. Although different types of additized gasoline have been available on the market for a long time, there is no reference method for analyzing the detergent-dispersant additives they contain.

Figure 14 illustrates the structure of some of the additives available on the market.

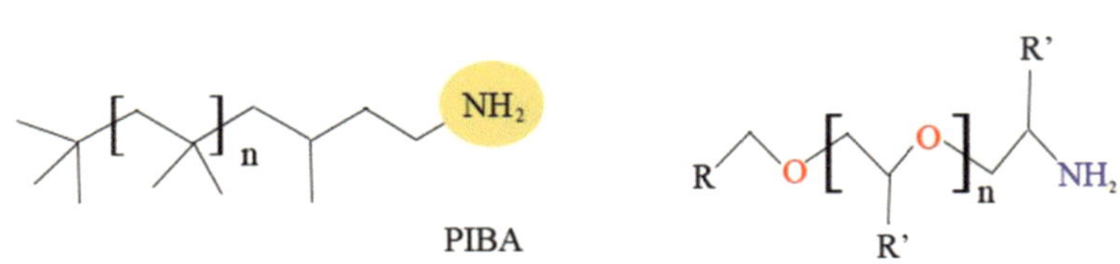

Figure 14.
Structure of some types of additives available on the market. Source: Santos [53].

One factor that hampers the development of analytical methods for detergent-dispersant additives is that they are sold as packages, with the additives already dissolved in solvents compatible with the hydrocarbon composition of the fuel, without indicating their respective composition.

Santos et al. have developed two laboratory methods for analyzing detergent-dispersant additives in gasoline. One uses thermogravimetry to identify the type of additive in the gasoline, which is then quantified by size-exclusion chromatography with refraction index detector, while the other just uses thermogravimetry [53–55].

d'Avila and Souza have developed a method that can be used in the field for the qualitative and/or semiquantitative identification of detergent-dispersant additives in fuels and lubricating oils. It is based on differences in the chromatographic behavior, in nonpolar stationary phases like silica, of nonpolar fuels and oils that form the basis for lubricants, which, due to their hydrocarbon composition in nonpolar stationary phases like silica, are retained less than detergent-dispersant additives, which are polar. The basicity of amine groups in additives could be used to reveal their position in the chromatography column by the addition of different acid-base indicators, making the presence or absence of the additives in the fuels and lubricating oils unequivocal. The process could be used by customers, inspectors, producers, or even by distributors—i.e., by operators who are not necessarily technically qualified—because of the simplicity of the procedure and the interpretation of the results, as shown in **Figure 15**, for a gasoline additized with ethanol, as is commonplace in Brazil and in many other countries [56]. The stages are:

1. Extraction of the ethanol from the gasoline with a saline aqueous solution, as described in Brazilian standard NBR13992, which is available and executable at all fuel distribution stations to easily determine the ethanol content in the gasoline

2. Continuous introduction of gasoline without ethanol in the column made of a Pasteur pipette containing a polar adsorbent (e.g., silica), using a hypodermic syringe

3. Addition of an alcoholic solution of different acid-base indicators, followed by the addition of ethanol to remove any excess, revealing the presence of additive in the column due to the "turning" of the indicator and its change of color because of the change in pH caused by the presence of the additive in the column

When the fuel contains a detergent-dispersant additive, a colored ring is formed when the indicator changes color (in **Figure 15** exemplified by bromothymol blue), enabling qualitative identification by the naked eye. Semiquantitative identification may be possible, as the intensity of the color is proportional to the concentration of the additive in the fuel.

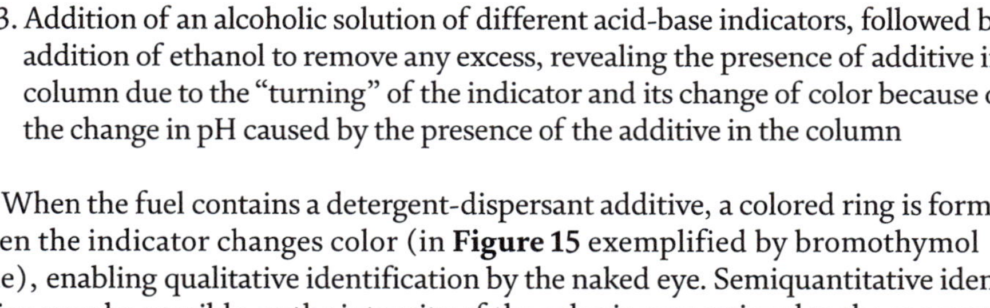

Figure 15.
Stages involved in the qualitative identification of detergent-dispersant additives in fuels, executable in the field. Source: d'Avila et al., 2012 (BR1020120292300/WO2014/075158A1) [56].

3.3.3 Adulteration

Lee et al. (2011) have developed a sensor to detect gasoline adulteration based on the solvatochromic behavior of a polydiacetylene (PDA) conjugated polymer embedded in matrix polymers of polystyrene and polyacrylic acid in the form of fibers. When exposed to pure and adulterated gasoline, they change to different colors and also change color in the presence of potential adulterants like paint thinner, methanol, and toluene (see **Figure 16**) [57].

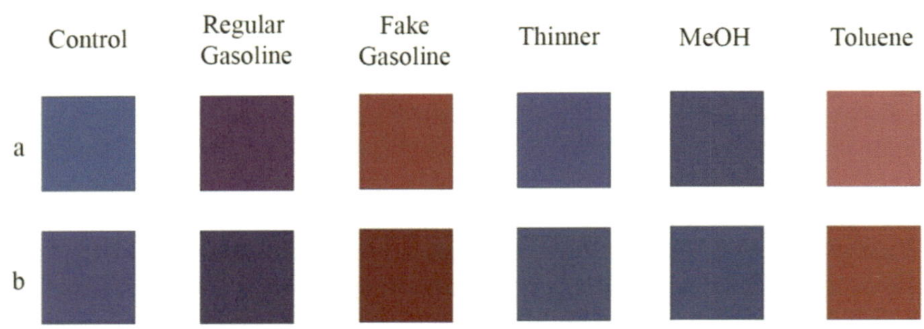

Figure 16.
Photos after the exposure of the PDA fibers in polystyrene to commercial and adulterated gasoline, thinner, methanol, and toluene on glass (a) and on silica TLC plate (b). Adapted from Lee et al. [57].

4. Conclusions

The quality and adulteration of fuels have been the object of several publications, especially with the use of modern analytical techniques, such as chromatography, spectroscopy, and chemometric methods, but with very few examples of colorimetric techniques. Among the colorimetric techniques shown here, developed for fuel quality and adulteration monitoring, especially those that are based on simple, rapid, low-cost procedures with potential for application in the field, we highlight those traditionally used for the qualitative identification of functional groups in organic analysis. The colorimetric methods for detecting biodiesel and vegetable oil adulterants in diesel-biodiesel blends are based on the qualitative analysis of carboxylic esters using the hydroxamic acid test, which produces a colored ferric hydroxamate complex. The colorimetric detection of methanol in ethanol and in gasoline was also based on a reaction for identifying aldehydes by the reaction for the formation of a Schiff base, which is colored. These assays are based on spot tests developed and described by Fritz Feigl and colleagues, whose publications [27, 48, 58] describe an incommensurable archive of spot tests used in organic and inorganic analyses, which could serve as inspiration for the development of other colorimetric assays. Likewise, the solvatochromic effect of dyes and fibers, used as probes or sensors, has been used to detect biodiesel in diesel-biodiesel blends and water or ethanol in gasoline and even for detecting adulterants in gasoline. Publications that use solvatochromic effects for the analysis of fuels based on undergraduate experiments—showing their operational simplicity—are also highlighted [43]. The solvatochromic effect has been the target of several studies by Christian Reichardt and colleagues and also represents an opportunity for the development of new colorimetric techniques [59–61].

Author details

Amanda Pereira Franco dos Santos[1*], Kissya Kropf da Silva[1], Gisele Alves Borges[1,2] and Luiz Antonio d'Avila[1,2]

1 Petroleum Products and Fuels Laboratory (LABCOM), School of Chemistry, Federal University of Rio de Janeiro, Rio de Janeiro, Brazil

2 Chemical and Biochemical Process Engineering Program, School of Chemistry, Federal University of Rio de Janeiro, Rio de Janeiro, Brazil

*Address all correspondence to: apfranco@eq.ufrj.br

References

[1] Wiedemann LSM, d'Avila LA, Azevedo DA. Adulteration detection of Brazilian gasoline samples by statistical analysis. Fuel. 2005;**84**(4):467-473. DOI: 10.1016/j.fuel.2004.09.013

[2] Vempatapu BP, Kanaujia PK. Monitoring petroleum fuel adulteration: A review of analytical methods. TrAC Trends in Analytical Chemistry. 2017;**92**:1-11. DOI: 10.1016/j.trac.2017.04.011

[3] Babua V, Krishnab R, Mani N. Review on the detection of adulteration in fuels through computational techniques. Materials Today: Proceedings. 2017;**4**: 1723-1729. DOI: 10.1016/j.matpr.2017.02.013

[4] ANP. Cartilha dos Postos Revendedores de Combustíveis [Internet]. 2017. Available at: http://www.anp.gov.br [Accessed: 07-01-2019]

[5] Kawano MS. Aplicação de transdutores óticos para a monitoração da qualidade de biocombustíveis [Application of optical transducers to fuel quality monitoring] [Doctoral thesis]. Curitiba: Universidade Tecnológica Federal do Paraná; 2016

[6] Wiedemann LSM, d'Avila LA, Azevedo DA. Brazilian gasoline quality: Study of adulteration by statistical analysis and gas chromatography. Journal of the American Chemical Society. 2005;**16**(2):139-146. DOI: 10.1590/S0103-50532005000200003

[7] Takeshita EV. Adulteração de gasolina por adição de solventes: análise dos parâmetros físico-químicos [The adulteration of gasoline by the addition of solvents: analysis of physicochemical parameters] [Master's dissertation]. Florianópolis: Universidade Federal de Santa Catarina; 2006

[8] Ranhotra SS. Checking automobile fuel adulteration using image processing techniques. In: Proceedings of the IEEE Second International Conference on Image Information Processing (ICIIP '13); 9-11 December 2013; Shimla. New York: IEEE; 2014. pp. 592-596. DOI: 10.1109/ICIIP.2013.6707661

[9] Pedroso MP, Godoy LAF, Ferreira EC, Poppi RJ, Augusto F. Identification of gasoline adulteration using comprehensive two-dimensional gas chromatography combined to multivariate data processing. Fuel. 2008;**1201**:176-182. DOI: 10.1016/j.chroma.2008.05.092

[10] Estadão. Carga Tributária Complexa Estimula Sonegação e Fraude no Setor de Combustíveis [Internet]. 2018. Available at: https://economia.estadao.com.br/noticias/geral,carga-tributaria-complexa-estimula-sonegacao-e-fraude-no-setor-de-combustiveis,70002295324 [Accessed: 09-01-2019]

[11] ANP. Resolução ANP N° 40 de 25 de Outubro de 2013 (DOU 28 de Outubro de 2013; Republicada DOU 30 de Outubro de 2013) [Internet]. 2013. Available at: http://www.anp.gov.br [Accessed: 16-08-2018]

[12] ANP. Boletim de Monitoramento da Qualidade dos Combustíveis [Internet]. Dezembro de 2018. Available at: http://www.anp.gov.br [Accessed: 18-12-2018]

[13] Mabood F, Gilani SA, Albroumi M, Alameri S, Nabhani MMOA, Jabeen F, et al. Detection and estimation of Super premium 95 gasoline adulteration with Premium 91 gasoline using new NIR spectroscopy combined with multivariate methods. Fuel. 2017;**197**:388-396. DOI: 10.1016/j.fuel.2017.02.041

[14] Cunha DA, Montes LF, Castro EVR, Barbosa LL. NMR in the time domain: A new methodology to detect

adulteration of diesel oil with kerosene. Fuel. 2016;**166**:79-85. DOI: 10.1016/j.fuel.2015.10.078

[15] Bhowmik S, Panua R, Ghosh SK, Debroy D, Paul AA. Comparative study of artificial intelligence based models to predict performance and emission characteristics of a single cylinder diesel engine fueled with diesosenol. Journal of Thermal Science and Engineering Applications. 2018;**10**(4):041004/1-041004/11. DOI: 10.1115/1.4038709

[16] The Star. Drastic Drop in Fuel Adulteration After ERC Tightens Grip on Culprits [Internet]. 2018. Available at: https://www.the-star.co.ke/news/2018/07/18/drastic-drop-in-fuel-adulteration-after-erc-tightens-grip-on-culprits_c1788907 [Accessed: 23-01-2019]

[17] Stratas Advisors. Regional Fuel-Quality Issues in Asia-Pacific [Internet]. 2016. Available at: https://stratasadvisors.com/api/GeneratePdf?url=http%3A%2F%2Fstratasadvisors.com%2Fen%2FInsights%2F090116-Asia-Pacific-Fuel&title=Regional%20Fuel-Quality%20Issues%20in%20AsiaPacific&secure=False [Accessed: 13-01-2019]

[18] Gaydou V, Kister J, Dupuy N. Evaluation of multiblock NIR/MIR PLS predictive models to detect adulteration of diesel/biodiesel blends by vegetal oil. Chemometrics and Intelligent Laboratory Systems. 2011;**106**(2):190-197. DOI: 10.1016/j.chemolab.2010.05.002

[19] Australian Government. International fuel quality standards and their implications for Australian Standards [Internet]. 2014. Available at: http://www.environment.gov.au/protection/publications/international-fuel-quality-standards [Accessed: 19-12-2018]

[20] Mazivila SJ. Trends of non-destructive analytical methods for identification of biodiesel feedstock in diesel-biodiesel blend according to European Commission Directive 2012/0288/EC and detecting diesel-biodiesel blend adulteration: A brief review. Talanta. 2018;**180**:239-247. DOI: 10.1016/j.talanta.2017.12.057

[21] Brito LR. Determinação de aditivos detergentes e dispersantes em gasolina utilizando a técnica do *ring oven* e imagens hiperespectrais na região do infravermelho próximo [Determination of detergent-dispersant additives in gasoline using the ring oven technique and near-infrared hyperspectral imaging] [Master's dissertation]. Recife: Universidade Federal de Pernambuco; 2014

[22] ASTM 1500/2012. Standard Test Method for ASTM Color of Petroleum Products (ASTM Color Scale)

[23] Corgozinho CNC, Carvalho MMO, PJS B. Spectrophotometric and Chemometric Determination of the ASTM Color of Automotive Diesel. Energy & Fuels. 2009;**23**(2):2136-2142. DOI: 10.1021/ef801019s

[24] NBR14359/2013. Produtos de Petróleo e Biodiesel-Determinação da Corrosividade-Método da Lâmina de Cobre

[25] ASTM 130/2018. Standard Test Method for Corrosiveness to Copper from Petroleum Products by Copper Strip Test

[26] DIN EN ISO 2160/1998. Petroleum Products—Corrosiveness to Copper—Copper Strip Test

[27] Feigl F. Spot Tests in Organic Analysis. 6th ed. Elsevier; 1960. pp. 249-250

[28] Costa Neto C. Análise Orgânica—Métodos e Procedimentos para Caracterização de Organoquímios. Rio de Janeiro: Editora UFRJ; 2004. p. 510

[29] Silva MAA, Correa RA, Tavares MO, NRA F. A new spectrophotometric

method for determination of biodiesel content in biodiesel/diesel blends. Fuel. 2015;**143**:16-20. DOI: 10.1016/j.fuel.2014.10.048

[30] Santos APF. Desenvolvimento de métodos para separação e identificação do biodiesel e/ou óleo vegetal no óleo diesel e de aditivos em gasolina [Development of methods for separation and identification of biodiesel and/or vegetable oil in diesel and of additives in gasoline] [Master's dissertation]. Rio de Janeiro: Universidade Federal do Rio de Janeiro; 2012

[31] Silva MAA, Borges ATD, Filho NRA. Desenvolvimento de ensaio para análise qualitativa de biodiesel em misturas diesel-biodiesel para aplicação em postos revendedores de combustíveis. Química Nova. 2016;**39**(1):100-103. DOI: 10.5935/0100-4042.20150165

[32] Leite DSS, Fernandes TES. Desenvolvimento de método para separação e identificação de biodiesel e/ou óleos vegetais em BX [Development of methods for separation and identification of biodiesel and/or vegetable oil in BX] [Monography]. Rio de Janeiro: Universidade federal do Rio de Janeiro; 2013

[33] EN 14078/2014. Gaseous and liquid fuels, lubricants and related products of petroleum, synthetic and biological origin—Test method for the determination of Fatty Acid Methyl Ester (FAME) content in diesel fuel or domestic heating fuel by mid infrared spectrometry

[34] Costa RPM. Desenvolvimento de método para determinação do teor de biodiesel em óleo diesel empregando teste colorimétrico e processamento de imagem [Improvement in the determination of biodiesel content in diesel fuel using colorimetric reaction and digital image processing] [Master's dissertation]. Rio de Janeiro: Universidade Federal do Rio de Janeiro; 2015

[35] Fong JK, Xue ZL. A dye-doped optical sensor for the detection of biodiesel in diesel. Chemical Communications. 2013;**49**:9015-9017. DOI: 10.1039/c3cc43958e

[36] Muniz RO, Martins SB, Honório G, Cunha JN, Souza CG, Andrade DF, et al. Total glycerol analysis in biodiesel samples using solid phase extraction coupled with enzymatic spectrophotometric determination. Royal Society of Chemistry. 2019;**11**:767-773. DOI: 10.1039/c8ay02736f

[37] Greenhill S. US 2004/0137546A1. Method for determination of free and combined glycerin in biodiesel. Filed: July 15, 2004

[38] Valdez HC, Amado RS, Souza FC, d'Elia E, Vieira EC. Free and total glycerol determination in biodiesel samples by enzymatic method with colorimetric detection. Química Nova. 2012;**35**(3):601-607. DOI: 10.1590/S0100-40422012000300028

[39] Chaar RN, Airoldi ML, Turkovics F, Serralvo RN, Muniz R, Martins S, et al. FR1851142. Procede d'analyse de glycerol total d'un echantillon de biodiesel base sur la detection par spectrophotometrie. Filed: February 12, 2018

[40] Serralvo Neto R, Chaar RN, Turkovics F, d'Elia E, d'Avila, Gimenes C, et al. FR1872032. Procede d'analyse du glycerol libre dans un echantillon de biodiesel par voie enzymatique. Filed: November 29, 2018

[41] Budag R, Giusti LA, Machado VG, Machado C. Quality analysis of automotive fuel using solvatochromic probes. Fuel. 2007;**85**:1494-1497. DOI: 10.1016/j.fuel.2005.12.023

[42] El Seoud OA, Loffredo C, Galgano PD, Sato BM, Reichardt C. Have biofuel, will travel: A challenging and colorful experiment for an undergraduate

spectroscopy course. Journal of Chemical Education. 2011;**88**:1293-1297. DOI: 10.1021/ed1009464

[43] Galgano PD, Loffredo C, Sato BM, Reichard C, El Seoud OA. Introducing education for sustainable development in the undergraduate laboratory: Quantitative analysis of bioethanol fuel and its blends with gasoline by using solvatochromic dyes. Chemistry Education Research and Practice. 2012;**13**:147-153. DOI: 10.1039/c1rp90061g

[44] NBR 16041/2015. Etanol combustível—Determinação dos teores de metanol e etanol por cromatografia gasosa

[45] Martins GBCM. Desenvolvimento de uma metodologia portátil para análise de metanol em etanol e detecção de aldeídos visando a utilização em combustíveis e outros produtos comerciais [Development of a portable methodology for the analysis of methanol in ethanol and detection of aldehydes for potential use on fuels and other commercial products] [Master's dissertation]. Brasília: Universidade de Brasília; 2012

[46] Martins GBC, Montenegro MA, Suarez PAZ. Kit colorimétrico para detecção de metanol em etanol combustível para o monitoramento da qualidade de combustíveis. Química Nova. 2015;**38**:280-284. DOI: 10.5935/0100-4042.20140305

[47] Suarez PAZ, Martins GBC. Br PI 10.2012.012197-2. Método de estabilização do reagente de Schiff em diversos veículos, reagente de Schiff imobilizado em matrizes sólidas, processo de impregnação desse reagente, método de determinação analítica de amostras baseado na utilização do reagente de Schiff estabilizado, kit para a determinação especialmente de metanol e formol em produtos comerciais e suas aplicações. Filed: May 22, 2012

[48] Feigl F. Spot Tests in Organic Analysis. 6th ed. Elsevier; 1960. pp. 223-224

[49] Costa Neto C. Análise Orgânica—Métodos e Procedimentos para Caracterização de Organoquímios. Rio de Janeiro: Editora UFRJ; 2004. pp. 510

[50] ANP. Resolução ANP N° 19 de 15 de Maio de 2015 (DOU 16 de Maio de 2015; Republicada DOU 20 de Maio de 2015) [Internet]. 2015. Available at: http://www.anp.gov.br [Accessed: 04-02-2019]

[51] Giordano GF, Ferreira DCM, Carvalho TR, Vieira LCS, Piazzetta MHO, Lima RS, et al. Portable platform for rapid and indirect photometric determination of water in ethanol fuel samples. Analytical Methods. 2014;**6**:9497-9502. DOI: 10.1039/C4AY02255F

[52] ASTM E203. Standard Test Method for Water Using Volumetric Karl Fischer Titration

[53] Santos APF. Quantificação de aditivos detergentes e dispersantes em gasolinas por análise térmica e cromatografia [Quantification of detergent-dispersant additives in gasoline by thermal and chromatographic analysis] [Doctoral thesis]. Rio de Janeiro: Universidade Federal do Rio de Janeiro; 2018

[54] d'Avila LA, Dweck J, Santos APF, Silva KK. BR10201801540 Procedimento para análise quantitativa de aditivos detergentes/dispersantes em combustíveis líquidos ou em qualquer outro fluido hidrocarbônico por análise térmica. Filed: July 27, 2018

[55] Santos APF, Silva KK, Dweck J, d'Avila LA. Quantification of detergent-dispersant additives in gasoline by size-exclusion chromatography and thermogravimetry. Fuel. 2017;**194**: 166-170. DOI: 1 0.1016/j.fuel.2 017.01.009

[56] d'Avila LA, Souza CG, Chaar JS. BR1020120292300 (WO2014/075158A1). Processo para detecção de aditivos básicos em combustíveis e óleos lubrificantes e kit para realizar o referido processo. Filed: November 12, 2013 (May 22, 2014)

[57] Lee J, Balakrishnan S, Cho J, Jeon SH, Kim JM. Detection of adulterated gasoline using colorimetric organic microfibers. Royal Society of Chemistry. 2011;**21**:2648-2655. DOI: 10.1039/C0JM02287J

[58] Feigl F. Spot Tests Vol I: Inorganic Applications. 4th ed. Elsevier; 1954

[59] Reichardt C. Solvatochromic dyes as solvent polarity indicators. Chemical Reviews. 1994;**94**:2319-2358. DOI: 10.1021/cr00032a005

[60] Reichardt C. Pyridinium n-phenolate betaine dyes as empirical indicators of solvent polarity: Some new findings. Pure and Applied Chemistry. 2008;**80**:1415-1432. DOI: 10.1351/pac200880071415

[61] Reichardt C, Welton T. Solute solvent interactions and solvent effects on the position of homogeneous chemical equilibria. In: Solvents and Solvent Effects in Organic Chemistry. 4th ed. Wiley-VCH; 2010. pp. 7-64 and 107-163. DOI: 10.1002/9783527632220